REVISED THIRD EDITION

Texas Tech Atmospheric Science Group

An Introduction to
ATMOSPHERIC SCIENCE
lab manual

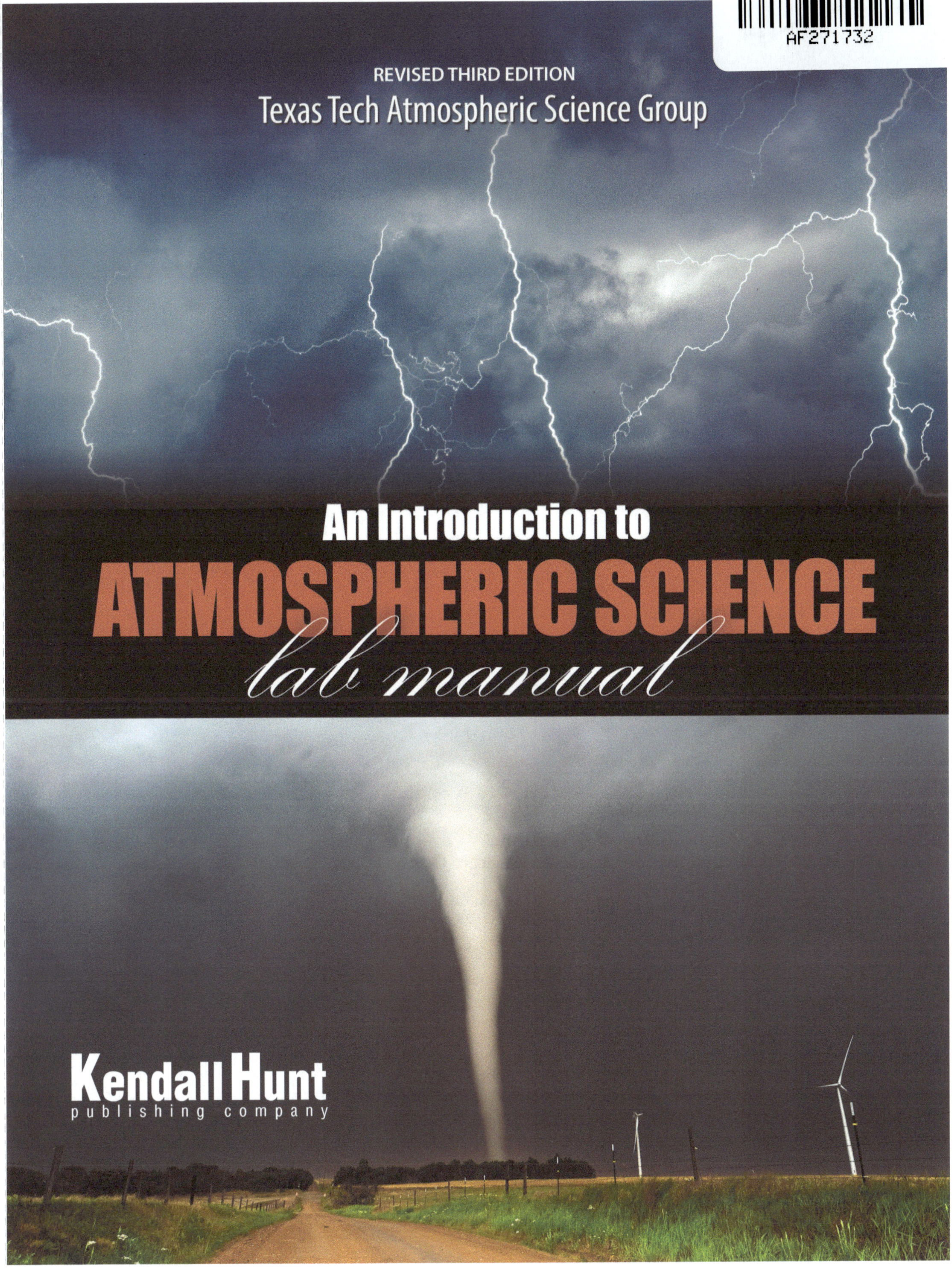

Kendall Hunt
publishing company

CONTENTS

PREFACE

This introductory level laboratory manual was written for the purpose of helping students achieve a basic knowledge and understanding of our atmosphere. The manual was designed to complement and reinforce a more thorough and comprehensive introductory atmospheric science course and textbook.

Throughout the lab manual important terms are highlighted in bold. A glossary is provided for easy reference.

Questions are interspersed throughout each of the eight labs. At the end of each lab, there is an answer sheet that also contains all of the questions. This is so the student can tear out the answer sheet and any accompanying maps for easy grading, and keep the background material and explanations for studying and future reference.

LAB 1

OBSERVING THE ATMOSPHERE

I. INTRODUCTION

A. OVERVIEW

One of the goals of atmospheric science, or meteorology, is to acquire new knowledge and gain a greater understanding of how the atmosphere works. One component of meeting this goal is observing the atmosphere. For example, researchers are investigating a type of thunderstorm called a **supercell**, in hopes of better understanding how these powerful storms produce tornadoes. To do this, researchers use Doppler radars to observe the winds within the thunderstorm. They also incorporate other instruments that measure temperature, humidity, wind, and pressure in close proximity to the storm.

You have probably heard of the *scientific method*. In atmospheric science, the scientific method is a way of investigating the atmosphere in order to acquire a greater knowledge and understanding of atmospheric phenomena.

The scientific method consists of four steps:
1. Identify a problem or question
2. Formulate a hypothesis
3. Collect and analyze data
4. Form a conclusion or theory

Collecting data in atmospheric science usually involves making observations.

Observing the atmosphere is very important in weather forecasting. Instruments are used to observe various weather elements such as temperature, pressure, wind, and humidity. These observations are made not only at the surface of Earth, but above the surface as well. They define the current state of the atmosphere. Forecasting is basically deciding how the current state of the atmosphere will change over a certain time period (e.g., hours, days, or weeks).

Modern day weather forecasters use high-speed computers to model atmospheric processes. These computers are programmed with mathematical equations that represent how weather variables such as temperature and wind will change over time. Weather observations from around the globe, both at the surface and in the atmosphere above the surface, are fed into these computers. The output is a forecast of precipitation, wind, and temperature among many, many others. There are limitations to the accuracy of these computer-generated forecasts. One limitation is the lack of observations in certain areas such as over the oceans.

All of us are observers of the atmosphere in one way or another. We see different types of clouds and precipitation. We see the blue color of the sky, a rainbow, or lightning. We hear thunder. In fact, it would not be possible to hear sounds were it not for the atmosphere's ability to propagate sound waves. We can feel the force of the wind. We feel the effects of high humidity on a hot summer day or the chilling effect of the wind on a cold winter day.

In this first lab, a few basic elements of weather and the instruments used to observe them will be presented. In addition, we will describe a way of displaying and analyzing weather observations that will enable you to visualize and interpret what is happening in the atmosphere.

B. OBJECTIVES

Upon completion of this lab, you should be able to:
1. Identify basic weather variables and explain how each is measured.
2. Interpret the symbols used to represent wind direction and speed on weather maps.
3. Draw isopleths on a weather map given plotted values of a particular weather element.

II. WEATHER VARIABLES

The state of the atmosphere at any given time and place is defined by weather elements such as temperature, pressure, wind, and humidity. Since these elements change in both space and time, we can call them *weather variables*. In this section, we will describe some of these variables and explain how they are measured.

A. TEMPERATURE

In scientific terms, **temperature** is a measure of the average kinetic energy of the molecules of a substance. **Kinetic energy** is the energy an object possesses because of its motion.

The liquid-in-glass thermometer (Figure 1) is the most common type of thermometer for measuring temperature. As the gas molecules in the atmosphere move at various speeds, they collide with the thermometer. When this occurs, energy is transferred to the thermometer. The faster the average speed of the gas molecules, the more energy is transferred. This causes the liquid (usually mercury) in the bulb of the thermometer to expand and move up the tube.

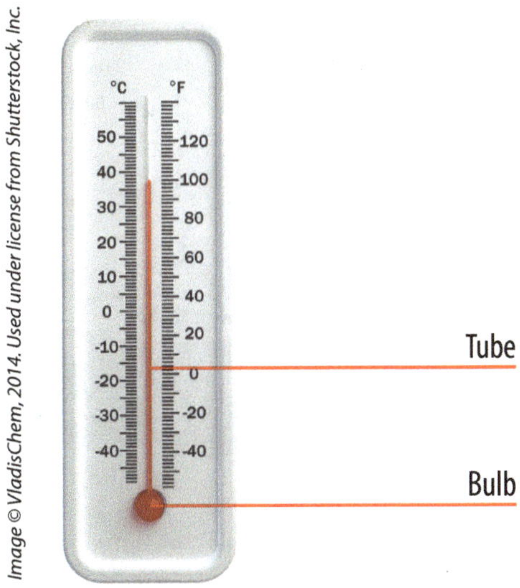

°C
°F
50
40
30
20
10
0
-10
-20
-30
-40
120
100
80
60
40
20
0
-20
-40
Tube
Bulb
Image © VladisChem, 2014. Used under license from Shutterstock, Inc.

FIGURE 1
Liquid-in-glass Thermometer

To obtain an accurate air temperature reading, a thermometer should never be exposed to direct sunlight. A thermometer in direct sunlight will absorb radiant energy from the sun, yielding a temperature reading higher than the actual air temperature.

Another type of thermometer, called a *thermistor*, indicates the temperature by measuring the resistance of a wire to the flow of electricity. The electrical resistance changes as temperature changes. As the temperature increases, the resistance decreases.

Temperature can be measured using the Fahrenheit scale (°F), the Celsius scale (°C), or the Kelvin scale (K). In the United States, temperatures measured at the surface of Earth are reported in °F. Temperatures measured above the surface are reported in °C. We'll explain how temperatures are measured above the surface of Earth in Lab 2. The Kelvin temperature scale is primarily used in scientific calculations. Formulas for converting from one temperature scale to another are in Appendix A.

B. PRESSURE

Pressure is defined as force per unit area. Weight is the force on an object due to gravity. So pressure can be thought of as a measure of the weight of the air above you.

Atmospheric pressure is measured using an instrument called a *barometer*. One type of barometer is the mercury barometer. Figure 2 illustrates how the mercury barometer works.

FIGURE  2

Mercury Barometer

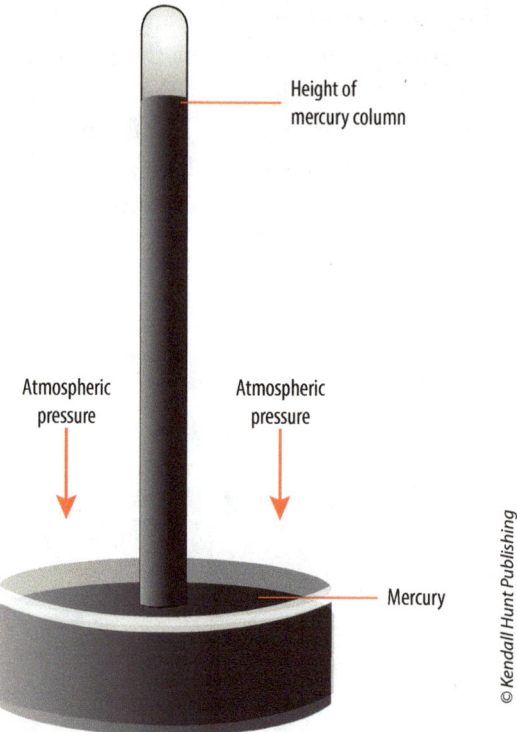

Height of
mercury column

Atmospheric
pressure

Atmospheric
pressure

Mercury

© Kendall Hunt Publishing

The weight of the atmosphere is balanced by the weight of the mercury in the glass tube. As atmospheric pressure changes, the height of the column of mercury changes. For example, lower atmospheric pressure (less weight) equals a lower height of the mercury column. The height of the column is a measure of atmospheric pressure. You have probably heard the barometric pressure given as so many inches of mercury such as 29.92 inches or 30.24 inches.

A second and more common type of barometer is the aneroid barometer (Figure 3).

FIGURE 3

Aneroid Barometer

Image © Dani Simmonds, 2014. Used under license from
Shutterstock, Inc.

Inside the barometer is a chamber or cell that is sealed. The cell expands or contracts as atmospheric pressure changes. That, in turn, moves the needle on the face of the instrument. For example, if the pressure decreases, the cell expands, which causes the needle to show a decrease in pressure.

Some aneroid barometers have descriptive weather terms such as *fair*, *stormy*, or *rain* printed on the face of the instrument. In general, it's the change in pressure over time that's important and not the actual value of pressure at one particular time. For example, if the pressure is falling, this might indicate the approach of a low pressure center. Low pressure centers are often associated with clouds and precipitation.

Airplanes are equipped with an aneroid barometer to indicate the plane's altitude. It will be shown in Lab 2 that as altitude increases, atmospheric pressure decreases (less weight above). The barometer is calibrated to read altitude instead of pressure. In this case, the instrument in the plane is called an *altimeter* (Figure 4).

FIGURE 4

An Altimeter

Image © Jan Kaliciak, 2014. Used under license from Shutterstock, Inc.

A unit of pressure used extensively by meteorologists is the **millibar**, abbreviated mb. A bar is a unit of pressure in the metric system. In a standard, or average atmosphere, the pressure at sea level is 1013.25 mb, which is equal to 29.92 inches of mercury.

Another unit of pressure that is gaining popularity is the hectopascal (hPa).

$$1 \text{ hPa} = 1 \text{ mb}$$

In this lab manual, we will use millibar for the unit of pressure.

C. DENSITY AND THE IDEAL GAS LAW

Another important variable in atmospheric science is **density**. Density is defined as mass per unit volume. There is no convenient instrument used to measure density in the atmosphere. It can be calculated if the air temperature and pressure are known. The relationship among temperature, pressure, and density is given by the following equation: $P = \rho R T$, where P is pressure, ρ is density, T is temperature, and R is a constant. This relationship is called *the ideal gas law* and is used to help explain the behavior of the atmosphere.

D. WIND

Wind is air in motion relative to the Earth's surface. Wind consists of a magnitude (speed) and a direction. It is important to know that the wind direction is given according to the direction <u>from</u> which the air is coming. For example, a south wind means the air is flowing from the south toward the north. A northeast wind means the air is flowing from the northeast.

An instrument used to measure wind speed is the *cup anemometer*. That's the one on the right in Figure 5. The movement of air causes the cups to spin. The number of rotations over time can be used to calculate wind speed. On the left side in Figure 5 is a *wind vane* used to measure wind direction.

FIGURE 5

Cup Anemometer and Wind Vane

Image © T.W. van Urk, 2014. Used under license from Shutterstock, Inc.

A more modern type of wind-measuring instrument is the *sonic wind sensor* (Figure 6). It uses sound waves to measure wind velocity.

Sonic wind sensor

FIGURE 6

Sonic Wind Sensor

Image courtesy of Steven Cobb

Humidity, a term associated with water vapor in the air, is another important weather variable. Humidity will be discussed in Lab 5.

Doppler radar and weather satellites are valuable observational tools. We will discuss Doppler radar in Lab 8 in conjunction with thunderstorms. Weather satellites provide images depicting the development, dissipation, and movement of clouds. This is especially significant where surface-based observations are lacking, such as over the oceans. Weather forecasters rely on satellite images to observe the development and track the movement of hurricanes. Satellite images can also be used to monitor the development of thunderstorms. An example of one type of satellite image called the *visible image*, can be seen in Appendix C. The image is based on the amount of sunlight reflected from clouds and the Earth's surface.

QUESTIONS

Note: You should answer questions in this lab and in all subsequent labs on the answer sheet found at the end of each lab.

1. If you were tasked with locating a liquid-in-glass thermometer outside to measure air temperature, which of the following would be the best location?
    A. On top of your car
    B. Attached to a west-facing wall of a building
    C. A place where the thermometer gets direct sunlight in the morning but not in the afternoon
    D. A place where the thermometer is shaded from the sun
2. What does an altimeter in an airplane measure to indicate the plane's altitude?

3. In a mercury barometer, what happens to the length of the mercury column in the glass tube as atmospheric pressure increases?
4. What is the average sea level pressure in millibars and inches of mercury?
5. The diagram on the answer sheet for this question shows directions (north, east, south, west) labeled with respect to point X. The arrows represent the wind flow relative to point X. How would a weather observer report the wind direction at point X?

III. SURFACE WEATHER MAPS

In the United States, observations of weather elements at the surface are routinely made by the Automated Surface Observing System, or ASOS. In addition to the weather variables previously discussed, ASOS measures other elements such as surface visibility, precipitation amounts, and the height of the base of clouds.

Another type of surface observing system is called a *mesonet*. It consists of densely spaced instruments that routinely measure weather variables.

Weather observations are recorded at numerous locations across the United States at the same time. These observations are then plotted on a map to obtain a "picture" of the weather across the country. In this section, we will introduce how wind speed and direction are plotted on a weather map. Then we will introduce the concept of isopleth analysis. In a later lab, we will show how other weather variables are plotted on weather maps.

A. PLOTTING WIND SPEED AND DIRECTION

Wind measurements are plotted on weather maps using symbols like the following:

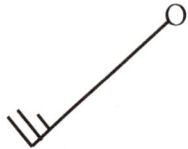

The circle represents the location where the observation was made. The longer line (shaft) extending from the circle represents wind direction. Imagine a compass with directions such as shown in the example below.

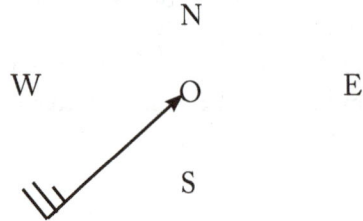

In this case, the wind direction is from the southwest. Wind direction is given *from* the direction the air is moving. It might be helpful to draw an arrow where the wind direction line meets the station circle (as shown above) to help visualize the flow of air.

Wind speed is shown by the shorter lines drawn at a 90° angle to the shaft. These lines are called "feathers." Wind speed is given in knots. A **knot** is a nautical mile per hour. Each full line coming off the wind direction line represents 10 knots. A half-line represents 5 knots.

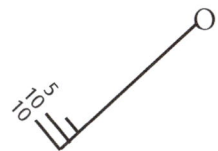

The wind speed is the sum of each wind speed line (feather). So in the example above, the wind speed is 25 knots.

Here are two other examples:

Wind from the northeast
at 15 knots

Wind from the east
at 10 knots

It is important to learn to read these wind symbols because they will be used in subsequent labs.

B. ISOPLETH ANALYSIS

Figure 7 on page 11 is an example of a surface weather map on which **sea level pressure** in millibars has been plotted at several locations. We will explain sea level pressure in Lab 2. Once observations taken at the same time at many locations are plotted on a map, how can we interpret all the data? A procedure called *map analysis* is performed in which isopleths are drawn to represent some variable. An **isopleth** is a generic term for a line drawn on a map that connects equal values of a given variable. Lines that connect equal values of temperature are called **isotherms**. Lines of equal pressure are called **isobars**.

Isopleths are drawn on weather maps to help us visualize and identify important meteorological features. Isobars help in identifying high and low pressure centers as well as evaluate pressure gradients that are related to winds. Isotherms can help identify the location of fronts. Basically, isopleths on weather maps reveal or enhance characteristics of the atmosphere contained within the observations.

When drawing isopleths, some basic rules must be followed:

a. The lines cannot cross or branch.
b. They are drawn using a fixed interval. For example, isobars on a surface weather map are typically drawn every 4 mb.
c. Lines are labeled with the value of that line.
d. Lines should be relatively smooth.
e. A particular line representing a certain value should be drawn so that plotted values on one side of the line are less than, and on the other side of the line are greater than the value of the line being drawn.

Use Figure 8 on page 13 to answer questions 6 and 7. This map shows isotherms that were drawn based on temperatures recorded at many surface locations at the same time.

6. Notice that the 60°F isotherm and the 70°F isotherm intersect at locations A and B. Think about what an isotherm represents and state why the intersection of those two isotherms does not truly represent the temperature at those locations at the time observations were made.

7. State one more violation of the rules for drawing isopleths (in this case isotherms) that can be seen on this map in Figure 8.

Figure 9 on page 14 shows temperatures in degrees Fahrenheit plotted at several locations. We'll use this figure to explain how isopleths (in this case isotherms) are drawn. The 70°F isotherm has already been drawn for you.

To understand why the 70°F isotherm was drawn this way, start on the lower right where the 70°F isotherm begins. There are no plotted values equal to 70°F on the lower right of the figure. This means we must interpolate between two plotted values. It is reasonable to assume there would be an observed temperature of 70°F between the 67°F and 73°F plotted temperatures. So the 70°F isotherm is drawn halfway between those two locations.

Continuing to follow the 70°F isotherm to the left, we find a similar situation where we must interpolate. Again, we can reasonably assume that a value of 70°F lies somewhere between the 66°F and 75°F. Continuing to the left, the 70°F isotherm must be drawn through the station that has a temperature of 70°F. The isotherm is then drawn just above the location with a temperature of 72°F since 70°F is closer to 72°F than 65°F. Note that all values above the 70°F isotherm are less than 70°F and values below the line are greater than 70°F.

8. On Figure 9, draw and label the 60°F and the 50°F isotherm.

Now, return to Figure 7 and analyze the sea level pressures by drawing isobars. The 1000 mb isobar and the 1020 mb isobar have already been drawn for you.

9. On Figure 7, draw and label isobars at 4 mb intervals starting with the 1004 mb isobar.

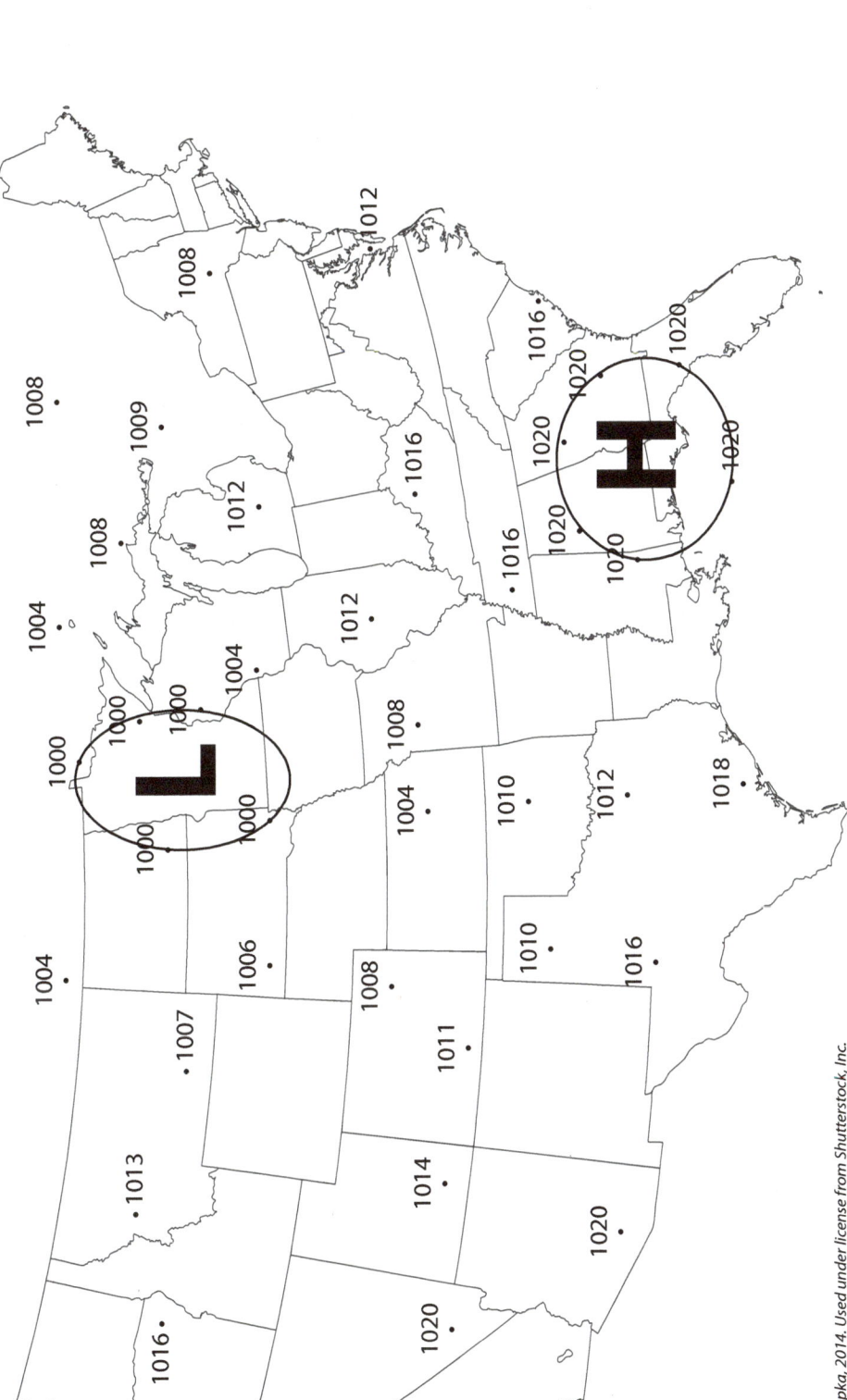

Map outline image © chrupka, 2014. Used under license from Shutterstock, Inc.

There is a relationship between horizontal changes in pressure and wind. In Lab 4, we will be using maps on which isobars have been drawn to explain this relationship. For now, it's important to learn some of the features revealed in the isobar pattern.

Notice in Figure 7 that there are two areas on the map with a closed isobar. One is in the north central part of the U.S. (the 1000 mb isobar) and the other is in the southeastern U.S. (the 1020 mb isobar). Having one or more closed isobars identifies either a low pressure center or a high pressure center. A low pressure center, also known as a **cyclone**, is located in the north central U.S. It is identified by the letter "L." Sea level pressures inside the closed 1000 mb isobar are less than 1000 mb. There is some location within this 1000 mb isobar that has the lowest pressure. If you started at that location and moved away in any direction, the pressure values increase.

In the southeast U.S. there is a high pressure center, also known as an **anticyclone**, identified by the letter "H." The location with the highest sea level pressure is inside the 1020 mb isobar. If you were to start at that location and move away in any direction, the pressure values decrease.

High and low pressure centers are important because they are associated with either rising or sinking air. Vertical motion is one of the major factors in determining the type of weather at a particular location.

Isobars may be spaced close together in some areas and farther apart in other areas. The isobar spacing has an important meaning that will be explained in Lab 4.

FIGURE 8

Isotherms Labeled in °F

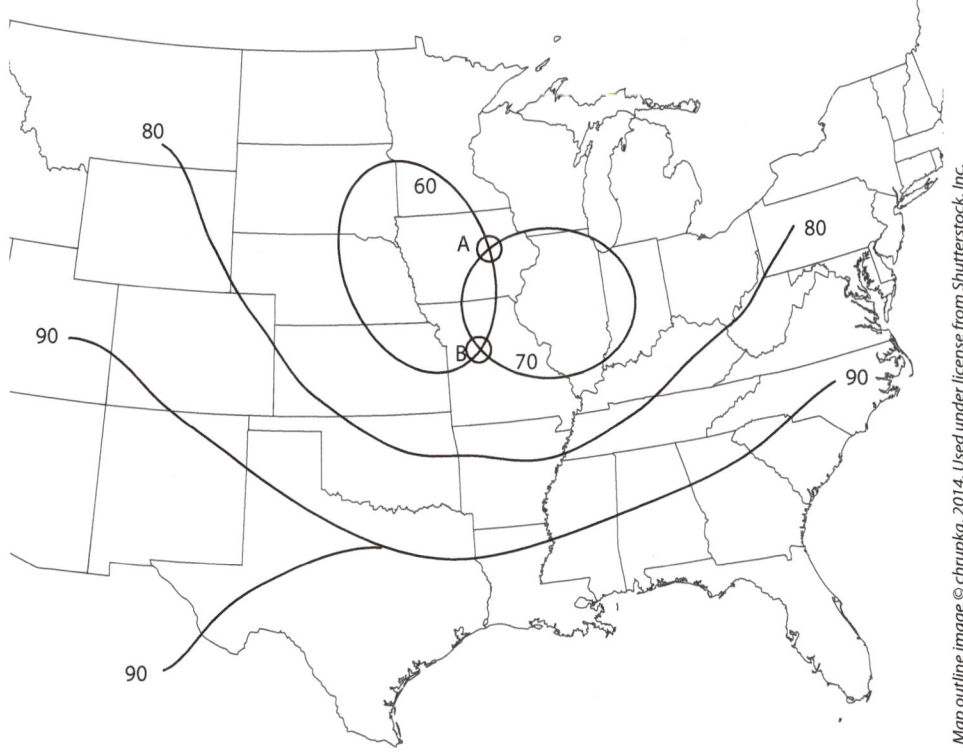

Map outline image © chrupka, 2014. Used under license from Shutterstock, Inc.

FIGURE  9

*A Plot of Surface
Temperatures*

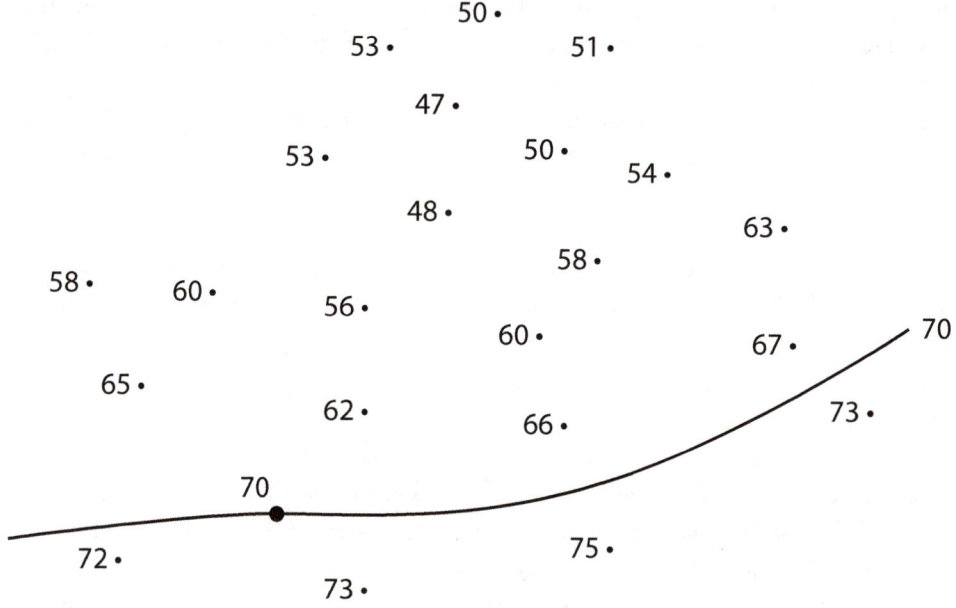

Name: _____

Course Number _____ Section Number _____

ANSWER SHEET FOR LAB 1

1. If you were tasked with locating a liquid-in-glass thermometer outside to measure air temperature, which of the following would be the best location? (Circle the correct answer.)
   A. On top of your car
   B. Attached to a west-facing wall of a building
   C. A place where the thermometer gets direct sunlight in the morning but not in the afternoon.
   D. A place where the thermometer is shaded from the sun

2. What does an altimeter in an airplane measure to indicate the plane's altitude?

3. In a mercury barometer, what happens to the length of the mercury column in the glass tube as atmospheric pressure increases?

4. What is the average sea level pressure in millibars and inches of mercury?

5. The diagram below shows directions (north, east, south, west) labeled with respect to point X. The arrows represent the wind flow relative to point X. How would a weather observer report the wind direction at point X?

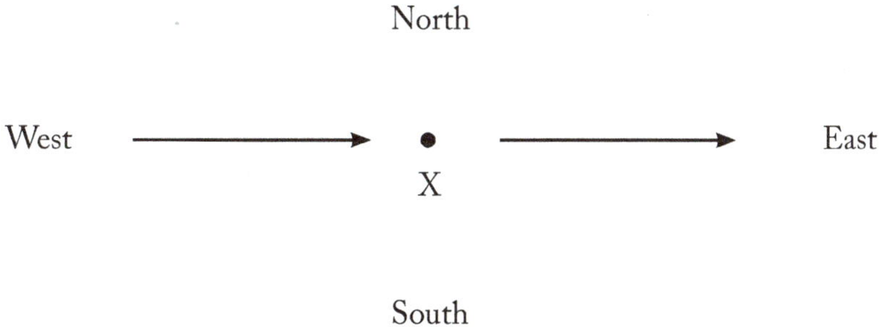

6. In Figure 8, notice that two different isotherms intersect at locations A and B. Think about what an isotherm represents and state below why the intersection of those two isotherms does not truly represent the temperature at those locations at the time observations were made.

7. State one more violation of the rules for drawing isopleths (in this case isotherms) that can be seen in Figure 8.

8. On Figure 9, draw and label the 60° and the 50° isotherm.

9. On Figure 7, draw and label isobars at 4 mb intervals starting with the 1004 mb isobar.

LAB 2

THE ATMOSPHERE IN 3-D

I. INTRODUCTION

A. OVERVIEW

To develop a basic understanding of how the atmosphere works, it is beneficial to think of the atmosphere that surrounds Earth as a three-dimensional (3-D) fluid. The atmosphere is a mixture of invisible gases. Gases, like liquids, are considered fluids. The weather we experience at the surface of Earth is, in large part, due to changes occurring in the atmosphere above the surface.

One example of how the atmosphere can be viewed as a 3-D fluid involves atmospheric pressure. In Lab 1, you were presented with a map of the U.S. that showed sea level pressure at numerous locations. The isobars you drew on this map graphically show the horizontal variation of pressure across the U.S. Pressure is a measure of the weight of the air above a certain location. The horizontal variation of pressure is due to changes in the amount of mass above each location. So there is a connection between the different values of sea level pressure observed over the U.S. and the vertical distribution of mass.

In this lab, we will primarily focus on the vertical structure of the atmosphere. We will discuss how temperature, pressure, and wind change in the vertical and how those variables are measured. An explanation of why sea level pressure is used on surface weather maps will be given. Lastly, we will introduce the term **jet stream**, which is part of the horizontal distribution of winds in the atmosphere. We will show how the jet stream is identified on weather maps and list some reasons why the jet stream is a significant feature in the atmosphere.

B. OBJECTIVES

Upon completion of this lab, you should be able to:
1. Explain how observations of certain weather variables are made above the surface.

17

2. Describe the vertical change of temperature in the atmosphere and identify the four layers of the atmosphere based on temperature.

3. Explain why sea level pressure rather than surface pressure is used on surface weather maps.

4. Define the term *jet stream* and identify its location using wind observations.

II. VERTICAL CHANGE OF TEMPERATURE

One way of describing the vertical structure of the atmosphere is by looking at how temperature changes with height (increasing altitude). Figure 1 shows the vertical temperature change in an average or reference atmosphere known as the U.S. Standard Atmosphere.

FIGURE 1

Temperatures and Layers of the U.S. Standard Atmosphere

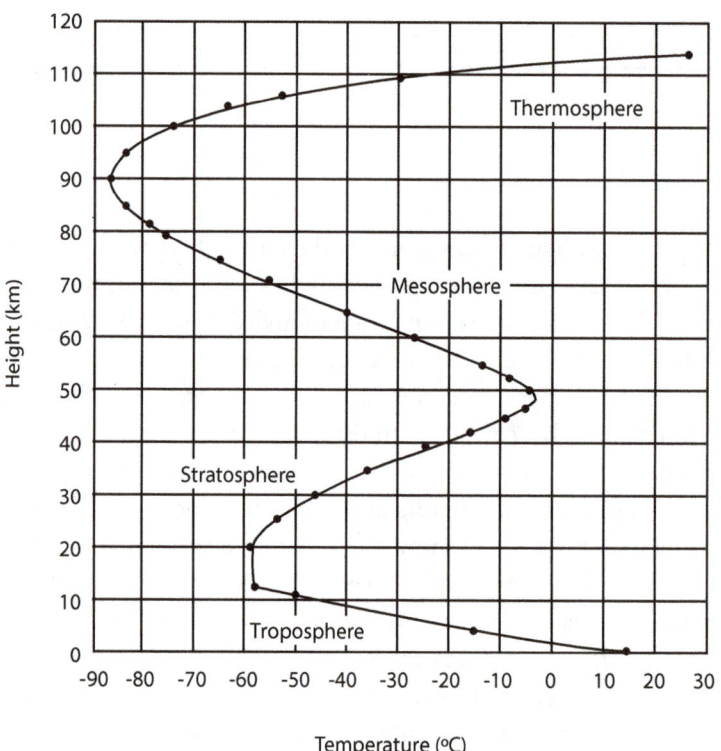

Notice in Figure 1 that there are layers of a certain depth in the atmosphere where temperature decreases with height and other layers where temperature increases with height. A decrease of temperature with height is called a **lapse rate.** An increase of temperature with height is called an **inversion.** This vertical variation in temperature makes it convenient to divide the atmosphere into 4 layers: troposphere, stratosphere, mesosphere, and thermosphere.

1. Using Figure 1, what layers in the atmosphere are characterized by a lapse rate?
2. What layers are characterized by an inversion?

In this lab, we will place more emphasis on the troposphere. The troposphere is where the vast majority of weather occurs.

The boundary between the troposphere and stratosphere is the **tropopause.** There is also a stratopause and a mesopause. In Figure 1, the tropopause is found at an altitude of 11 km. It marks the altitude where temperature stops decreasing with height.

The rate at which temperature changes with height in the troposphere has a profound effect on what kind of weather we observe. The type of cloud that forms and the type of precipitation (rain, snow, etc.) that falls to the ground depend on the vertical change of temperature.

A device called a **radiosonde** (Figure 2) is used to observe vertical variations in temperature and other variables through the troposphere into the stratosphere.

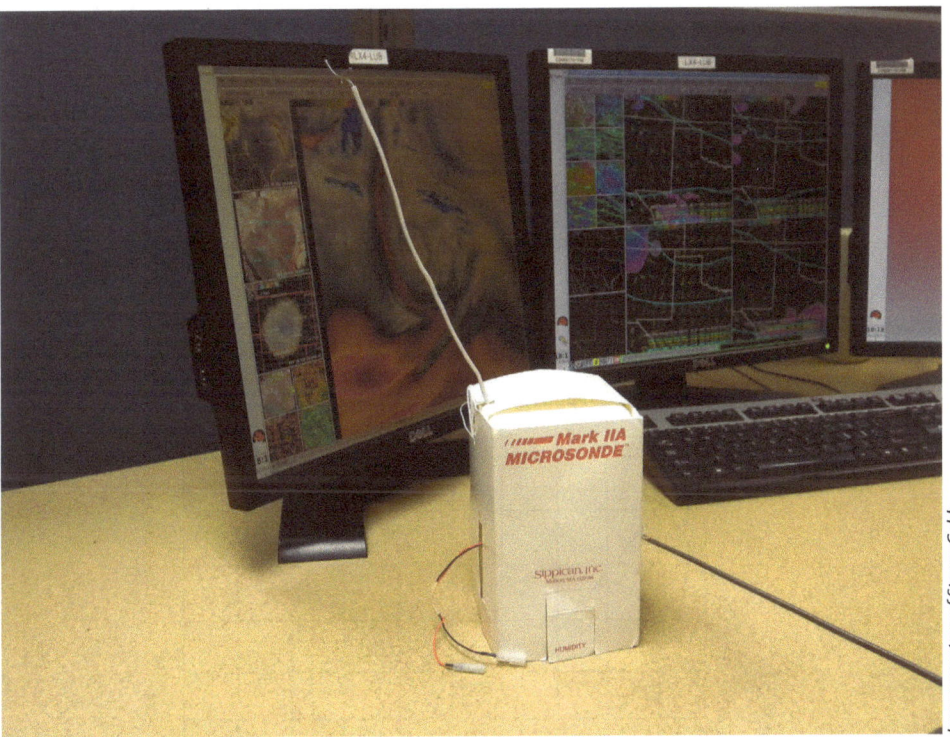

FIGURE 2

A Radiosonde

The radiosonde contains instruments that measure temperature, pressure, and humidity. It is attached to a balloon filled with either helium or hydrogen. Once the balloon is released from the surface, the balloon carries the radiosonde upward. Measurements of temperature, pressure, and humidity are made at numerous altitudes. A radio transmitter sends the data back to a computer on the ground. The radiosonde can be tracked using the Global Positioning System (GPS). This tracking allows the wind direction and speed to be computed at various altitudes. Since wind measurements can be obtained, the term *rawinsonde* is sometimes used.

Rawinsonde observations are made routinely twice a day, 12 hours apart, at selected locations across the U.S. Figure 3 shows the locations in the 48 adjoining states. A rawinsonde observation is called an upper air observation or a sounding.

FIGURE 3

Rawinsonde Observation Sites

Figure 4 illustrates how temperatures in the vertical can vary from one geographic location to another.

FIGURE 4

Temperature Soundings from Barrow, AK and Hilo, HI

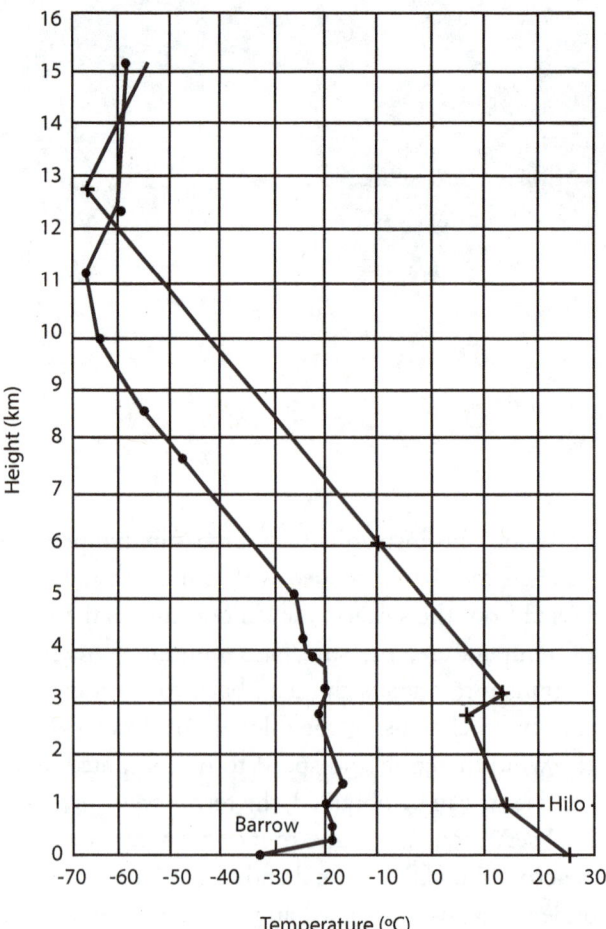

In this figure, temperatures measured by a radiosonde for two locations have been plotted. The dots connected by a solid line are for Barrow, Alaska, in the Arctic. The plus signs connected by a solid line are for Hilo, Hawaii, in the tropics.

3. Which sounding (Barrow or Hilo) shows an inversion close to the surface?

An inversion near the surface is not uncommon, especially in the early morning. These inversions are usually shallow, only a few hundred meters in depth. This is because at night, Earth's surface cools which allows the air in contact with the surface to cool by **conduction**. Conduction is the transfer of energy from molecule to molecule within a substance or from one substance in contact with another. The air just above the surface doesn't cool as much.

4. On the Hilo sounding, notice there is an inversion in the lower part of the troposphere. This inversion is between what two altitudes?

These examples show that shallow inversion layers can be observed within the troposphere.

5. Using Figure 4, what is the height of the tropopause to the nearest km at each location?

    Barrow: _____

    Hilo: _____

The tropopause is not at the same height everywhere over the Earth. The height of the tropopause is greatest at the equator and lowers as you go toward the poles. The height of the tropopause, or stated another way, the depth of the troposphere, is related to the average temperature of the troposphere at some geographical location.

6. In Figure 4, compare the temperatures at Barrow and Hilo at several altitudes in the troposphere. Without doing any calculation, does it appear that the average temperature in the troposphere at Barrow is warmer or colder than that at Hilo?

7. Based on your answers to questions 5 and 6, complete this statement: As the average, or mean, temperature in the troposphere increases, the height of the tropopause _____.
    (fill in the blank with either *increases* or *decreases*)

III. VERTICAL CHANGE OF PRESSURE

The vertical structure of the atmosphere can also be described by the way pressure changes with height. Figure 5 depicts this change for the U.S. Standard Atmosphere. The curved line shows what the pressure is at various altitudes in millibars (mb). For example, the pressure at 1 km in the Standard Atmosphere is 900 mb.

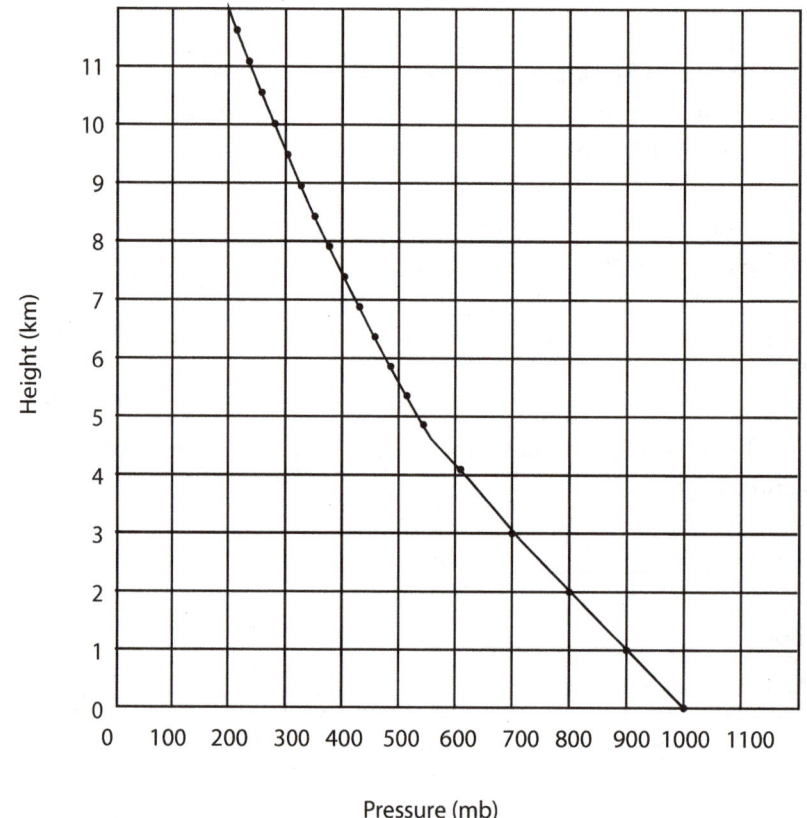

In this figure, we are just looking at the troposphere. Atmospheric pressure continues to decrease upward above the troposphere at an ever slower rate.

The average sea level pressure is 1013.25 millibars. This value represents a certain amount of mass in the atmosphere that causes a weight or force per unit area. Half of the average sea level pressure is about 500 mb.

8. Using Figure 5, locate 500 mb on the bottom horizontal line. Move up the vertical grid line until it intersects the pressure curve that has been drawn on the graph. What is the altitude where 500 mb is found?

9. Using the conversion formulas in Appendix A, convert this altitude to both miles and feet.

Since pressure can be used to represent the distribution of mass in the atmosphere, you should realize that one-half the mass of the entire atmosphere lies between sea level and the altitude you determined in question 9. This demonstrates how concentrated, or dense, the atmosphere is in the lower part of the troposphere.

Commercial aircraft generally fly between 30,000 and 40,000 feet above sea level. Let's use 35,000 feet, which is approximately 11 km. In the U.S. Standard Atmosphere, this is the height of the tropopause.

10. Using Figure 5, what is the pressure in mb at 11 km in the Standard Atmosphere?

The value you should have obtained in question 10 is much lower than the average sea level pressure. The air in the upper troposphere is very thin, meaning the density (mass per unit volume) of the air is very low. Such a low density means the pressure is also going to be very low. The low density of the air explains why the cabin in commercial aircraft is pressurized and why the flight attendants explain how to use the oxygen masks should the cabin lose pressure.

Keep in mind that the theme of this lab is based on the atmosphere as a 3-D fluid. We are now going to see how the vertical change in pressure is related to horizontal changes in pressure.

As can be seen in Figure 5, pressure decreases with height quite rapidly in the lower portion of the troposphere.

11. Using Figure 5, calculate the change, or difference, in pressure between 0 km and 5 km (5 km is approximately 3 miles).

The number you should have obtained in question 11 is much larger than a typical pressure difference between the highest and lowest pressure over the entire U.S.! Again, pressures drop very dramatically as you move upward through the lower troposphere. This presents a problem if surface pressure measurements at many locations over the country were plotted on a weather map and analyzed by drawing isobars. **Surface pressure** (also called station pressure) is the pressure measured at a given location on Earth due to the weight of air above that location. If surface pressures were plotted on a weather map and analyzed, there would always be low pressure in mountain areas and high pressure along the coast. That would not be a true representation of differences in pressure from place to place caused by processes occurring in the atmosphere. Pressure differences from one location to another due to differences in elevation are greater than pressure differences due to changes occurring in the atmosphere.

The problem with using surface pressure arises because places where pressure is measured are not all at the same elevation above sea level. Since pressure decreases so dramatically in the vertical, locations at a high elevation, such as Denver, Colorado (elevation 5,280 feet above sea level), will always have a lower pressure compared to locations near sea level, such as Houston, Texas.

Figure 6 is a side view of the terrain showing five cities (A, B, C, D, and E) at different elevations with respect to sea level. Use this figure to answer question 12.

FIGURE 6

*Cities at Different
Elevations*

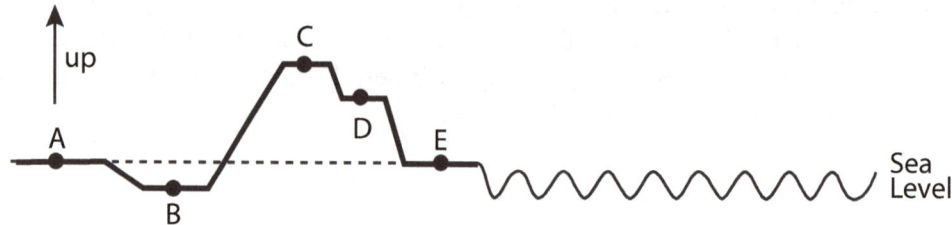

12. If the surface pressure was measured at the same time on a given day at each location, which location(s) would have the lowest surface pressure?

To eliminate the effect of altitude on the pressure observations, the surface pressure at each location is adjusted to a sea level pressure. This is done by accounting for the amount of mass in an imaginary layer of air between the surface and sea level. This amount of mass can be accounted for by the fact that on average, pressure in the lower atmosphere changes 10 mb for every 100 meter change in altitude. So if a city were 500 meters above sea level, 50 mb would need to be added to the surface pressure to yield a sea level pressure. For the rare location below sea level, the adjustment involves subtracting from the observed surface pressure. Sea level pressure is what the pressure would be if the location were at sea level.

For question 13, let's say city "C" in Figure 6 represents Denver and city "E" represents Galveston, Texas (approximately at sea level).

13. a. Denver's elevation is 1600 meters. On a particular day, the surface pressure at Denver is 852 mb. What is the sea level pressure at Denver?

On this same day, the sea level pressure at Galveston is 1008 mb.

b. What is the surface pressure at Galveston? (assume Galveston is at sea level)

c. Which city has the higher surface pressure?

d. Which city has the higher sea level pressure?

Converting surface pressure to sea level pressure removes the elevation difference among locations. Therefore, any differences in pressure from one place to another will be due to meteorological processes occurring in the atmosphere. This explains why *sea level* pressures are plotted on surface weather maps.

IV. VERTICAL CHANGE OF WIND

Both wind speed and direction can change as you go from the surface upward through the troposphere. In general, wind speeds increase with height in the troposphere. The highest wind speeds are typically found close to the tropopause.

As stated earlier, wind measurements at various altitudes in the troposphere can be obtained by tracking the radiosonde as it is carried upward by a balloon. Wind speed and direction can be plotted on a diagram such as seen in Figure 7. These data were obtained from a rawinsonde observation at Denver, Colorado.

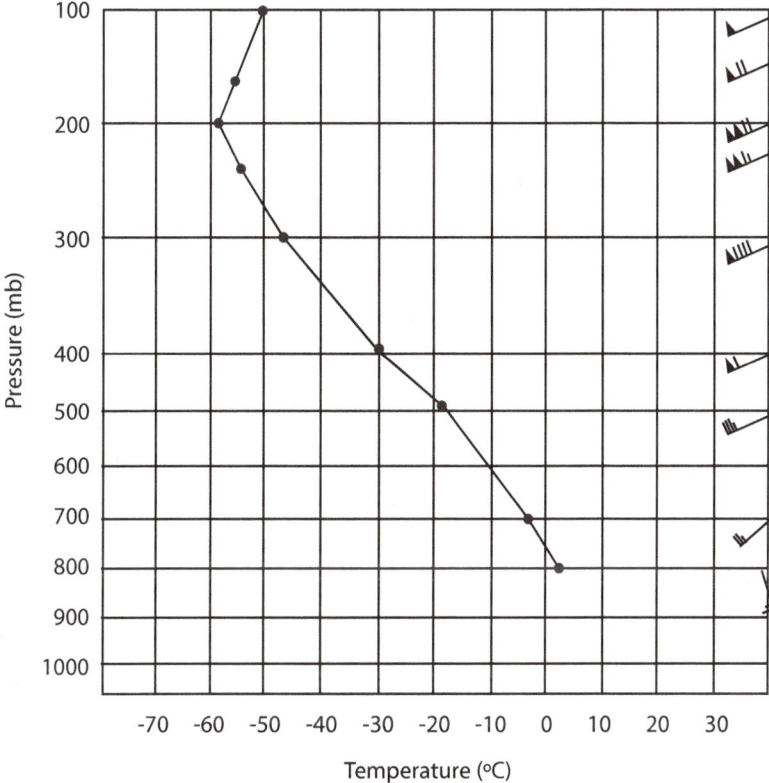

FIGURE 7

Sounding at Denver, Colorado

Notice that pressure values, in millibars, are used on the left vertical axis in Figure 7 to represent height. Since pressure decreases with height, we can use pressure to represent height instead of feet or meters. This is useful because the radiosonde measures pressure.

Wind measurements at different levels are plotted on the right vertical axis in Figure 7. Wind refers to the horizontal movement of air. So the wind symbols plotted in Figure 7 show the change in horizontal wind speed and direction in the vertical.

The symbols used to plot wind direction and speed are the same type of symbols used to plot winds on weather maps as discussed in Lab 1. For example, in Figure 7 the wind at 500 mb is from the southwest at 35 knots.

Note how the wind at 300 mb is plotted in Figure 7:

A shaded triangle represents 50 knots. Each of the other lines represents 10 knots. Adding up all the lines gives a wind speed of 90 knots. The wind direction is from the southwest.

Temperature measurements taken at various altitudes by the radiosonde launched at Denver are shown by the solid line in Figure 7.

14. Based on the way temperature changes with height at Denver, what is the height of the tropopause in terms of pressure?

15. Looking at the change in wind speed with height in Figure 7, what is the highest wind speed plotted and what is the pressure where the highest wind speed occurs?

You should discover that the highest wind speeds are found near the tropopause.

Since rawinsonde observations are taken at many locations across the U.S., we can plot wind speed and direction measured at a specific altitude (or pressure) for all of those locations. This will give a picture of how winds change in the horizontal.

Figure 8 is a plot of wind observations at approximately 34,000 feet over the U.S. Since the average height of the tropopause is about 36,000 feet, we can say this map represents winds near the tropopause.

FIGURE 8

Wind Observations at Approximately 34,000 Feet

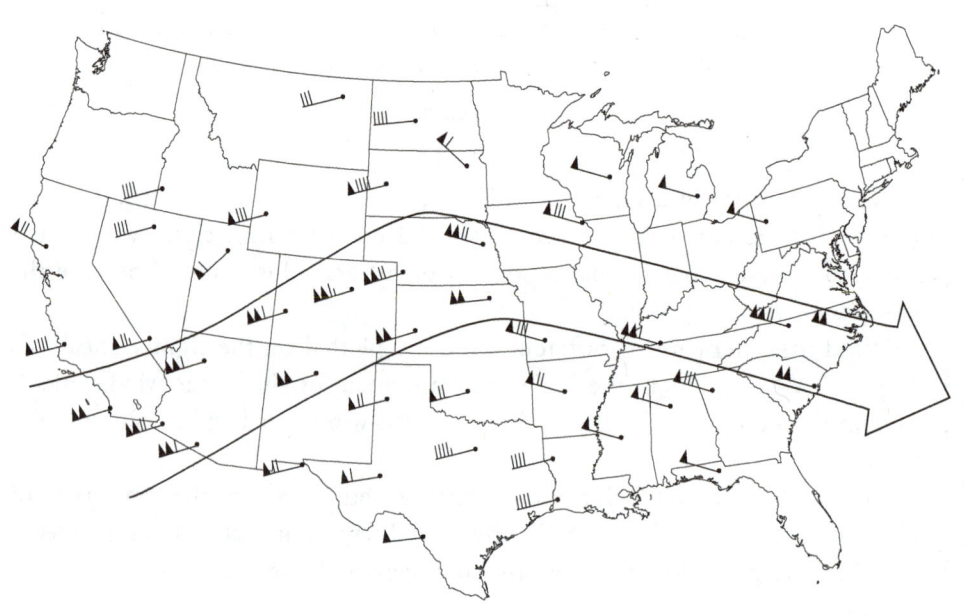

Wind speed and direction are plotted using the same type of symbols discussed earlier. The large arrow drawn on the map in Figure 8 represents a very important feature in the atmosphere called the *jet stream*. The jet stream is a relatively narrow band of very fast moving air found near the tropopause.

16. Notice the wind speeds within the jet stream in Figure 8. As you move north or south of the jet stream, do wind speeds increase or decrease?

17. In general, are winds in the jet stream from an easterly direction or from a westerly direction?

The jet steam represents the core of the strongest winds in the upper troposphere.

18. Figure 9 on page 29 is a plot of observed winds similar to Figure 8 except for only a portion of the United States. On this figure, draw an arrow that identifies the jet stream over the western U.S.

The jet stream is important for several reasons. One is that it is associated with the development and movement of low pressure centers. The jet stream is often associated with the development of severe thunderstorms. It is important in aviation. Planes flying near the jet stream will have either a strong head wind or a tail wind depending on the direction of flight. In addition, turbulence is sometimes encountered when flying near the jet stream.

FIGURE 9

Wind Observations at Approximately 34,000 Feet

Map outline image © chrupka, 2014. Used under license from Shutterstock, Inc.

Name: _____

Course Number _____ Section Number _____

ANSWER SHEET FOR LAB 2

1. Using Figure 1, what layers in the atmosphere are characterized by a lapse rate?

2. What layers are characterized by an inversion?

3. In Figure 4, which sounding (Barrow or Hilo) shows an inversion close to the surface?

4. On the Hilo sounding in Figure 4, notice there is an inversion in the lower part of the troposphere. This inversion is between what two altitudes?

5. Using Figure 4, what is the height of the tropopause to the nearest km at each location?

   Barrow: _____

   Hilo: _____

6. In Figure 4, compare the temperatures at Barrow and Hilo at several altitudes in the troposphere. Without doing any calculation, does it appear that the average temperature in the troposphere at Barrow is warmer or colder than that at Hilo?

7. Based on your answers to questions 5 and 6, complete this statement: As the average, or mean, temperature in the troposphere increases, the height of the tropopause _____. (fill in the blank with either *increases* or *decreases*)

8. Using Figure 5, locate 500 mb on the bottom horizontal line. Move up the vertical grid line until it intersects the pressure curve that has been drawn on the graph. What is the altitude where 500 mb is found?

9. Using the conversion formulas in Appendix A, convert this altitude to both miles and feet.

10. Using Figure 5, what is the pressure in mb at 11 km in the Standard Atmosphere?

11. Using Figure 5, calculate the change, or difference, in pressure between 0 km and 5 km (5 km is approximately 3 miles).

12. Using Figure 6 on page 24, if the surface pressure was measured at the same time on a given day at each location, which location(s) would have the lowest surface pressure?

13. a. Denver's elevation is 1600 meters. On a particular day, the surface pressure at Denver is 852 mb. What is the sea level pressure at Denver?

On this same day, the sea level pressure at Galveston is 1008 mb.

b. What is the surface pressure at Galveston? (assume Galveston is at sea level)

c. Which city has the higher surface pressure?

d. Which city has the higher sea level pressure?

14. Based on the way temperature changes with height at Denver in Figure 7, what is the height of the tropopause in terms of pressure?

15. Looking at the change in wind speed with height in Figure 7, what is the highest wind speed plotted and what is the pressure where the highest wind speed occurs?

16. Notice the wind speeds within the jet stream in Figure 8. As you move north and south of the jet stream, do wind speeds increase or decrease?

17. In general, are winds in the jet stream from an easterly direction or from a westerly direction?

18. On Figure 9, draw an arrow that identifies a portion of the jet stream over the western U.S.

LAB 3

RADIATION AND TEMPERATURE

I. INTRODUCTION

A. OVERVIEW

Radiation is an energy transfer process. The sun emits radiation that is transferred to Earth. The surface of the Earth and objects on the surface, including you, radiate energy. Gases in the atmosphere selectively absorb and also radiate energy.

In this lab, we will discuss how the amount of solar radiation reaching Earth's surface changes during the year. This is the main cause of seasonal variations in temperature. We will also discuss radiation emitted by Earth's surface, as well as the important role of the atmospheric greenhouse effect in maintaining a livable world.

B. OBJECTIVES

Upon completion of this lab, you should be able to:
1. Explain how variations in solar angle result in variations in the amount of solar radiation (insolation) reaching Earth's surface.
2. Calculate the solar angle at a given latitude.
3. Describe the atmospheric greenhouse effect and explain why it is important.

II. THE ELECTROMAGNETIC SPECTRUM

Radiation can be described as energy that propagates in the form of electromagnetic waves. The waves have electric and magnetic properties. One way of categorizing a wave is by its **wavelength**. The wavelength is the distance between two subsequent wave crests (Figure 1).

FIGURE **1**

Wavelength

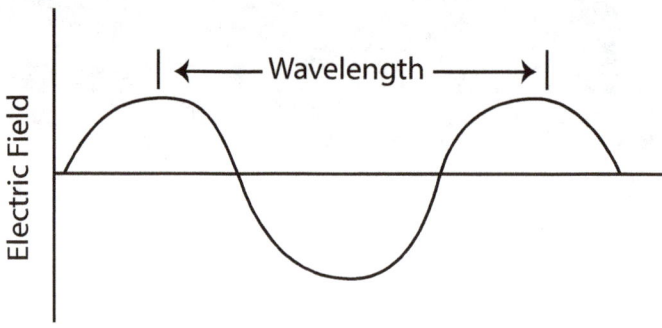

Radiation does not come in just one wavelength. There is a range of wavelengths called the *electromagnetic spectrum* (Figure 2).

FIGURE **2**

The Electromagnetic Spectrum

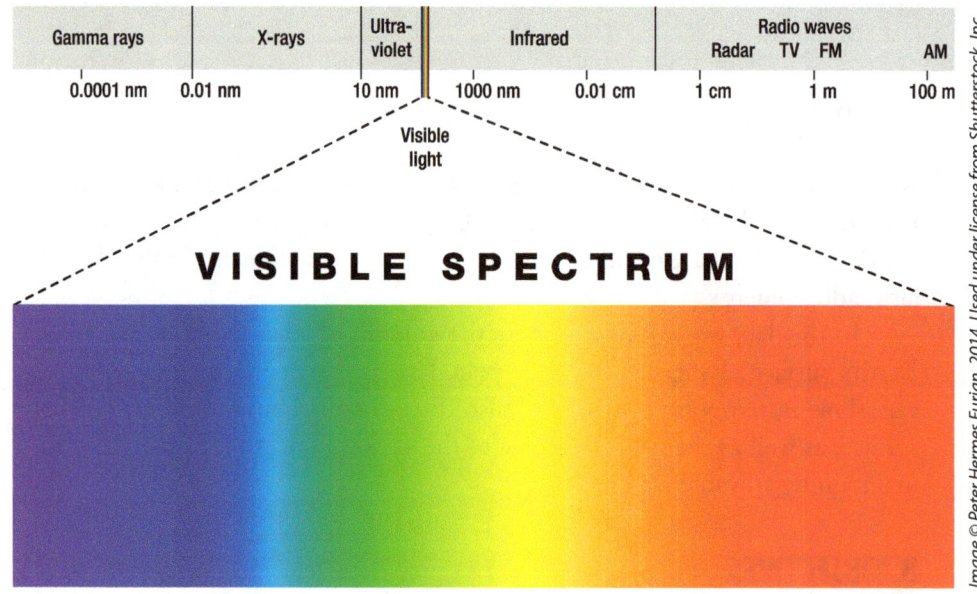

In this figure, notice that the shorter wavelengths are to the left and the longer wavelengths are to the right. It is useful to divide this continuous spectrum of wavelengths into sections and give them names such as ultraviolet radiation, visible radiation, and infrared radiation.

Two terms that are important in a discussion of radiation are **shortwave radiation** and **longwave radiation**. Shortwave radiation refers to energy emitted by the Sun. Longwave radiation refers to energy emitted by Earth and the atmosphere. The Sun radiates its maximum intensity in the visible part of the electromagnetic spectrum. Earth and the atmosphere radiate primarily in the infrared part of the spectrum. Infrared wavelengths are longer than visible wavelengths as can be seen in Figure 2. Hence, radiation from the Earth and atmosphere is called longwave radiation and radiation from the Sun is called shortwave radiation.

III. SOLAR RADIATION AND THE SEASONS

Radiant energy transferred from the Sun to Earth is the primary source of energy for all the processes that control our weather and climate. All points on the surface of Earth do not receive the same amount of solar energy during the year.

A. SEASONS

To understand why incoming solar radiation varies at a particular location during the year, one must first understand why Earth has seasons. Figure 3 shows the orientation of Earth with respect to the Sun at different times of the year. This is how an observer in space would view the orbit of Earth around the Sun.

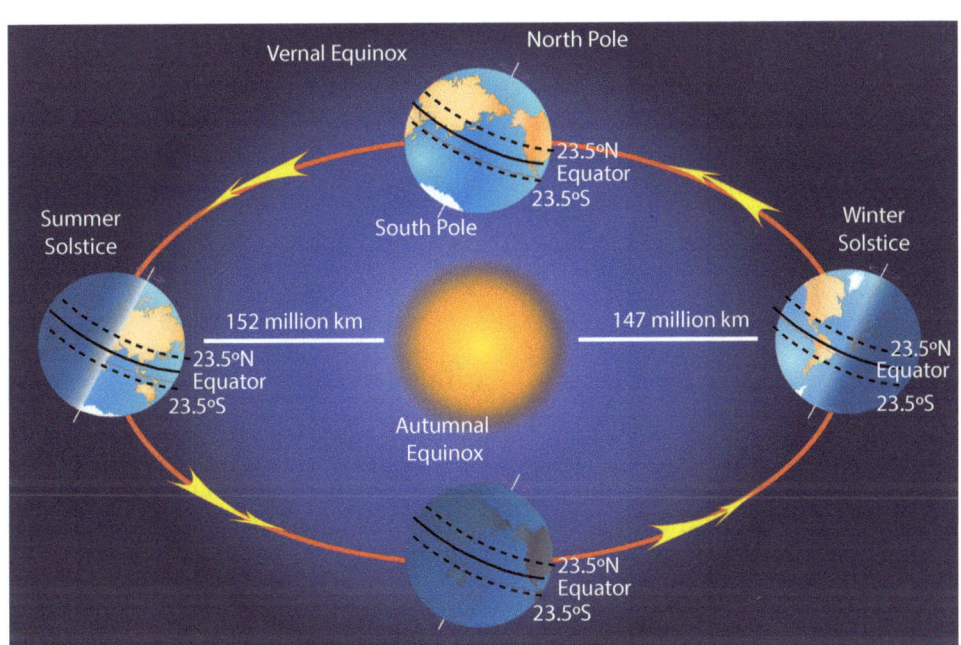

© Kendall Hunt Publishing

FIGURE 3

Orientation of Earth as it Orbits the Sun

Notice that the orbit is not a perfect circle. It is actually an elliptical orbit with Earth closest to the Sun in early January and farthest from the Sun in early July. However, the changing distance between Earth and the Sun is not the cause of seasons.

The reason that Earth has seasons is that the Earth's axis of rotation is tilted from the vertical by 23.5 degrees. The tilt of Earth's axis causes the amount of solar energy reaching Earth's surface to vary with latitude throughout the full year.

It is important to know the position of the Sun during the year with respect to specific latitudes shown in Figure 3. The latitude of 23.5 degrees North (23.5°N) is the northern most point on Earth where the Sun is directly overhead. This occurs on the Northern Hemisphere summer solstice (June 20/21). When the Sun is directly overhead at 23.5 degrees South latitude (23.5°S), the winter solstice occurs in the Northern Hemisphere (December 21/22). Both the vernal (spring) equinox

(March 20/21) and autumnal (fall) equinox (September 22/23) occur when the Sun is directly overhead at the equator. The Southern Hemisphere experiences the opposite season to the Northern Hemisphere. *In this lab, terms related to seasons such as summer or winter solstice pertain to the Northern Hemisphere.*

The different positions of the Sun with respect to these latitudes is due to the Earth maintaining a tilt of 23.5 degrees as it orbits the Sun. On any given day of the year, the Sun will be directly overhead at some latitude between 23.5°N and 23.5°S.

B. SOLAR ANGLE

The amount of radiant energy from the Sun received at any latitude on Earth during the year partially depends on the **solar angle** at that latitude. Solar angle is the angle between the horizon and the Sun. An example is shown in Figure 4.

FIGURE 4

Solar Angle

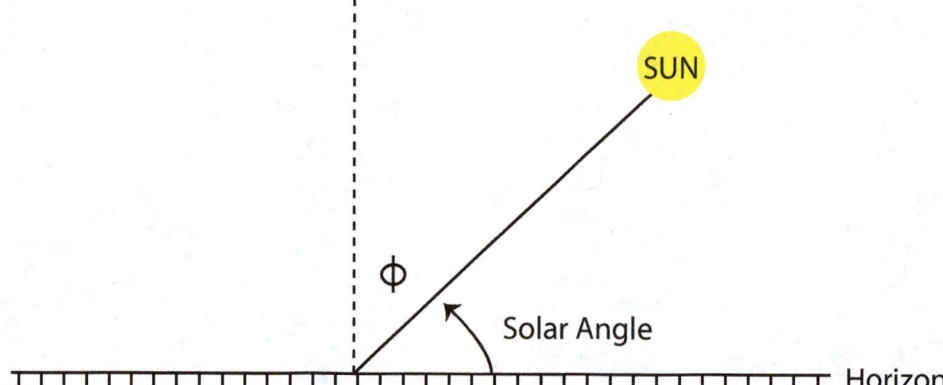

Solar angle is used to describe the Sun's position in the sky from the perspective of someone on Earth. A 90-degree solar angle means the Sun is directly overhead. An example would be at the equator on both the spring and fall equinox.

A corresponding term called **zenith angle** can also be used to describe the Sun's position in the sky as measured from vertical (ϕ in Figure 4). Zenith angle is the angle from a point directly overhead to the Sun. The solar angle and zenith angle added together equals 90 degrees. Solar angle will be used in this lab.

1. At what latitude is the solar angle 90° on the Northern Hemisphere winter solstice?
2. Assume a city is located at a latitude of 35°N. During the year, will the solar angle at that city ever be 90°? If you answer yes, during what month would a solar angle of 90° occur?

Solar angle can be computed for any given location. Since the position of the Sun in the sky changes between sunrise and sunset, solar angle is specified to be valid at solar noon. This is when the Sun is at its highest point in the sky and the solar angle is greatest. When comparing the solar angle at different locations, it is understood that solar angles are computed for solar noon.

To compute the solar angle for a given location on a particular day of the year, the **solar declination** must be known. Solar declination is the latitude at which the Sun appears directly overhead at solar noon. Its value will vary between +23.5° and -23.5°.

Table 1 shows how the solar declination varies during the year. Dates pertain to the Northern Hemisphere.

TABLE 1

Solar Declination

DATE	SOLAR DECLINATION
Vernal Equinox	0 degrees
Summer Solstice	+23.5 degrees
Autumnal Equinox	0 degrees
Winter Solstice	-23.5 degrees

For this lab, the following equation is used to compute the solar angle:

Solar Angle = 90° − (the latitude of a location) + (solar declination)

The solar declination will be a negative value if it is in the Southern Hemisphere.

3. Compute the solar angle at International Falls, Minnesota (INL) and Brownsville, Texas (BRO) for the dates given below. The latitude of International Falls is 48.6°N and the latitude of Brownsville is 25.9°N.

These two cities are identified on the map below for reference.

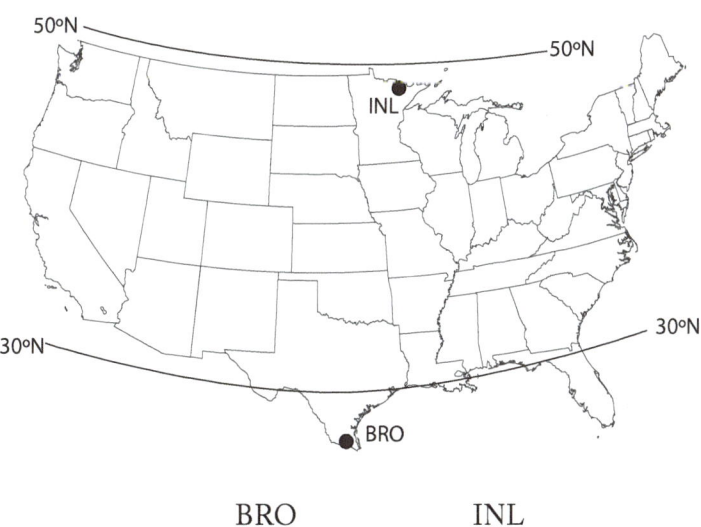

	BRO	INL
June 20/21	_____	_____
December 21/22	_____	_____

As you answer questions 4 through 7, try to visualize the Sun's position in the sky at solar noon at each location.

4. On the summer solstice, which location (BRO or INL) has the higher solar angle?

5. On the winter solstice, which location (BRO or INL) has the higher solar angle?

6. On both the summer and winter solstice, which location (BRO or INL) is closer to the latitude where the solar angle is 90°?

7. Assume it's the winter solstice and you are at the latitude where the solar angle is 90°. If you could compute the solar angle at each latitude as you move north to INL, would the solar angles increase or decrease?

Use the three diagrams below to answer questions 8 and 9. The diagrams depict different solar angles, similar to Figure 4.

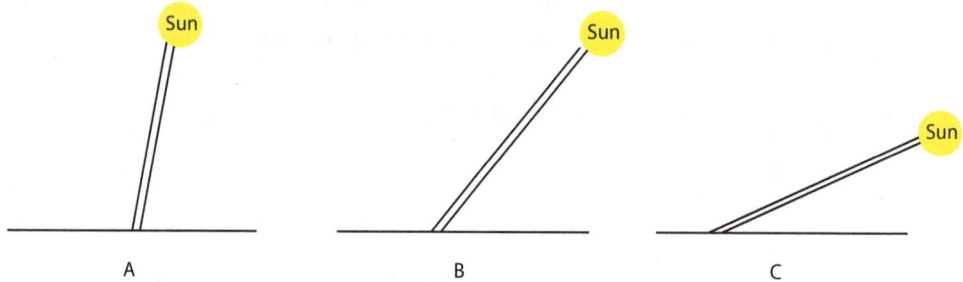

8. Which diagram (A, B, or C) best represents the Sun's position in the sky at solar noon at BRO on the summer solstice?

9. Which diagram (A, B, or C) best represents the Sun's position in the sky at solar noon at INL on the winter solstice?

C. SOLAR ANGLE, INSOLATION, & TEMPERATURE

Insolation refers to the amount of incoming solar energy during the daylight hours at some location. This daily insolation varies with latitude and time of year. Insolation at a particular location on Earth depends in large part on the solar angle. The amount of solar radiation received will affect how much heating occurs. Therefore, seasonal variations in temperature over Earth's surface are largely due to changes in the solar angle and not the varying distance of Earth from the Sun.

The higher the solar angle, the more insolation is received at a given location. The lower the solar angle, the less insolation is received. When the solar angle is high, the beam of radiant energy strikes Earth's surface more directly causing a greater amount of heating as opposed to a lower solar angle. This can be illustrated by shining a flashlight onto a horizontal surface at different angles (Figure 5).

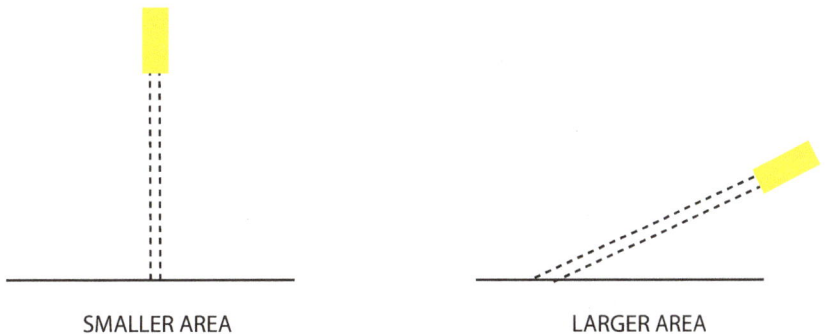

FIGURE   5

Flashlight Beam

SMALLER AREA   LARGER AREA

If the beam of light shines on the surface from a lower angle, the light spreads over a larger area. This means the intensity of light at some point will be less.

In addition, a lower solar angle means the radiant energy from the Sun must travel through more of the atmosphere compared to a higher solar angle before reaching Earth's surface (Figure 6).

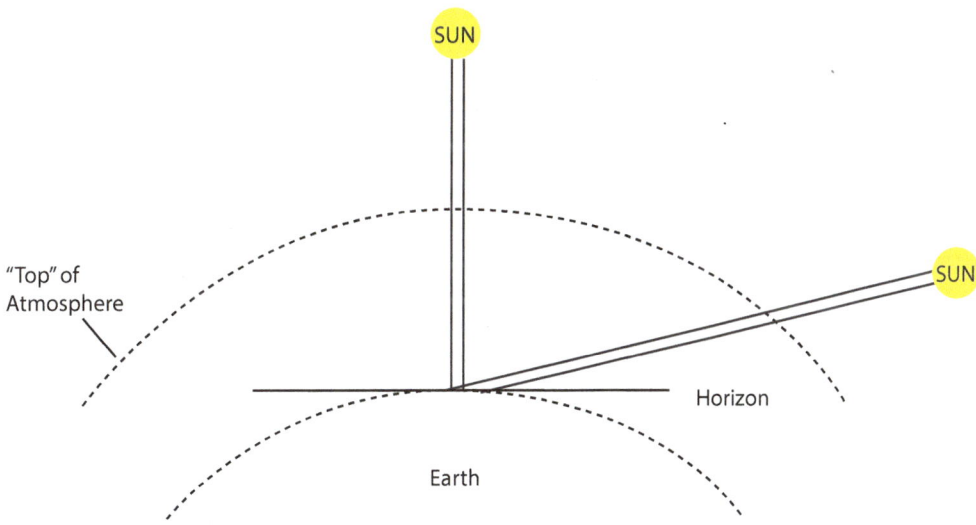

FIGURE   6

Path of Radiant Energy Through the Atmosphere

SUN

"Top" of Atmosphere

SUN

Horizon

Earth

Processes in the atmosphere such as **absorption** and **scattering** will reduce the amount of solar energy that arrives at Earth's surface. For a low solar angle, the beam of energy must travel through more atmosphere. So there will be a greater loss of radiant energy due to these processes than when the solar angle is higher.

Use the solar angles you computed for INL and BRO to answer questions 10 and 11.

10. At BRO, as you move forward in time from the winter solstice to the summer solstice:
    a. Does the solar angle increase or decrease?

    b. Does the amount of insolation increase or decrease?

11. On the winter solstice:
    a. Which location, BRO or INL, receives the lesser amount of insolation?

    b. Which location will have more surface heating by the Sun?

The temperature at any location on Earth is greatly influenced by the amount of insolation received at that location. The amount of insolation received is largely governed by the solar angle.

Figures 7a and 7b show average temperatures in January and July, respectively. Temperatures are in degrees Fahrenheit and have been adjusted to sea level. The solid lines drawn on the maps are isotherms (lines of equal temperature).

Map outline image © MaxFX, 2014. Used under license from Shutterstock, Inc.

FIGURE 7a.

Average Temperatures in January

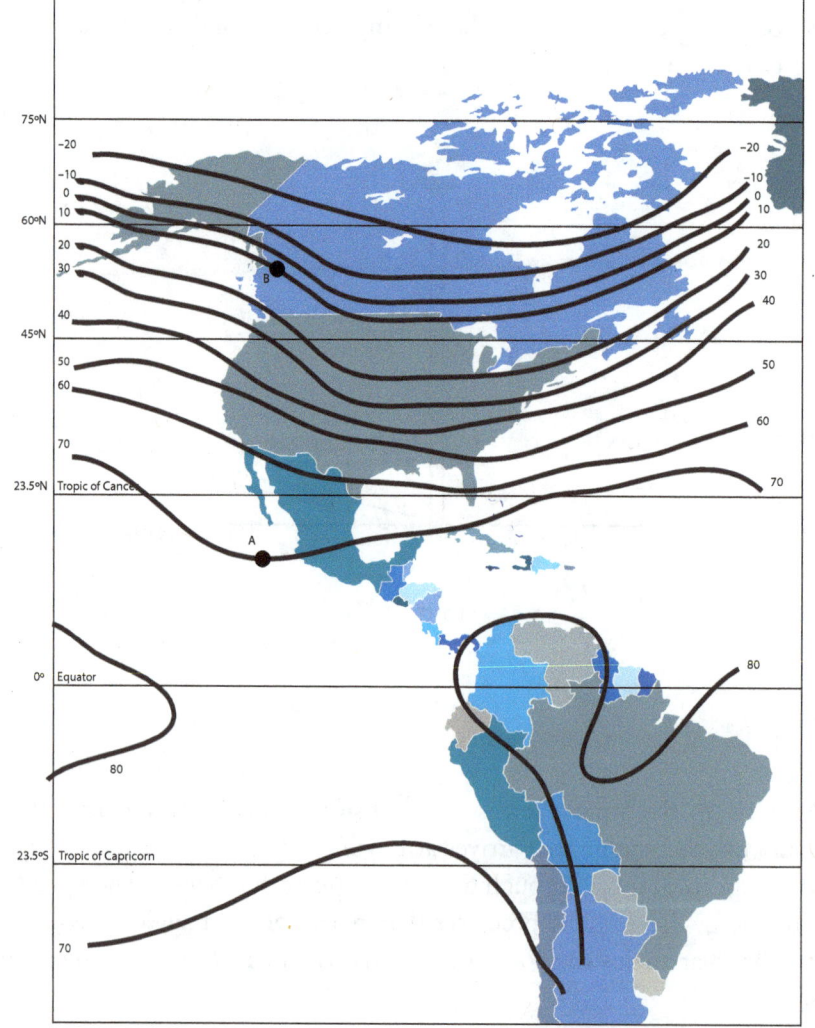

LAB 3 · *Radiation and Temperature*

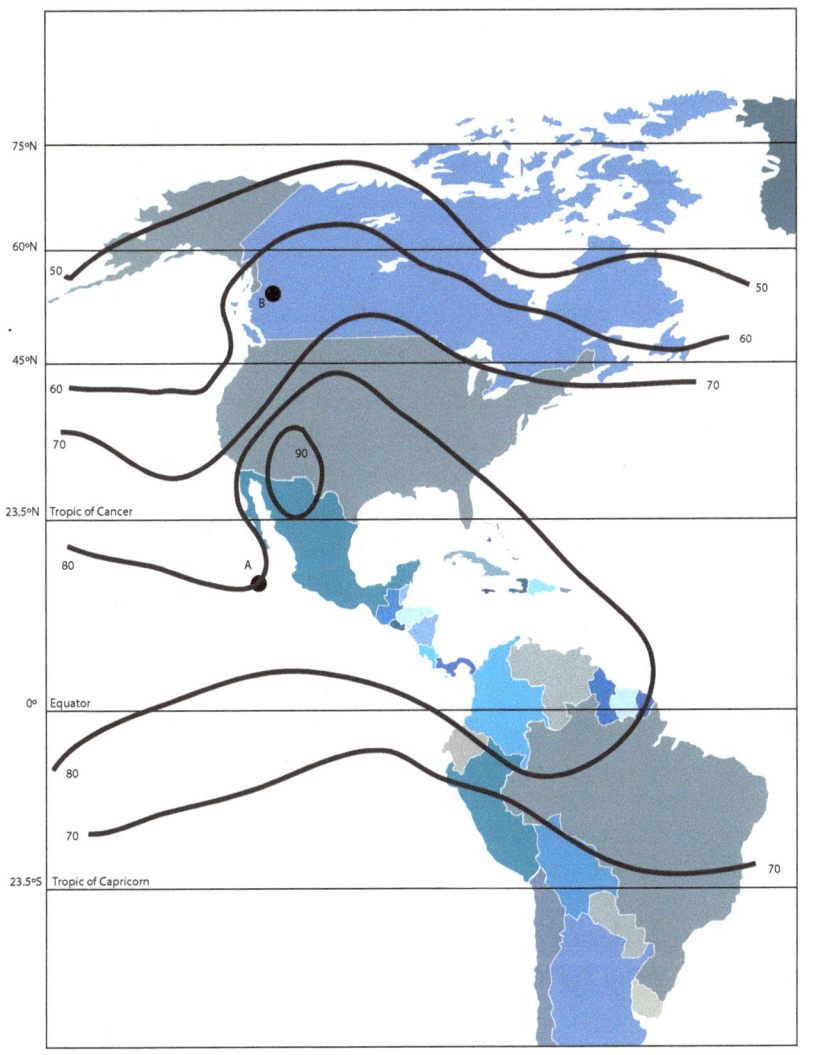

Map outline image © MaxFX, 2014. Used under license from Shutterstock, Inc.

Use these maps to answer question 12.

12. At point B, compare the average temperatures in January and July. The reason why the average temperature at point B in July is higher than the average temperature in January is largely because the solar angle at point B in July is _____ than it is in January.
(fill in the blank with either *higher* or *lower*)

You should be able to relate the temperatures in the Northern Hemisphere in January and July to the way Earth is tilted with respect to the Sun during the year as seen in Figure 3.

In addition to the solar angle, the length of day is also a factor controlling how much insolation reaches a particular location on Earth. The number of hours of daylight at each latitude affects the amount of insolation received. While this is an important component in fully understanding seasonal variations in temperature, we have chosen to emphasize solar angle in this lab.

IV. THE ATMOSPHERIC GREENHOUSE EFFECT

A. DESCRIPTION

The focus in section III of this lab was on solar radiation. Earth also emits radiation. Earth radiates virtually all of its energy at infrared wavelengths.

The *atmospheric greenhouse effect* involves the absorption and emission of infrared (longwave) radiation by certain gases in the atmosphere. These gases are known as greenhouse gases. Three of these gases are water vapor, carbon dioxide, and methane, with water vapor being the most abundant. Greenhouse gases absorb much of the infrared radiation emitted by Earth's surface. These gases radiate infrared radiation back to the Earth where it is absorbed. Figure 8 depicts this process.

FIGURE   8

Atmospheric Greenhouse Effect

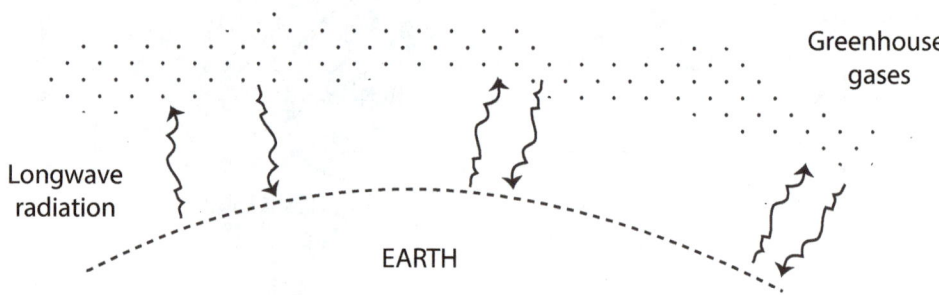

B. IMPORTANCE

When averaged over the entire surface of the Earth and for all seasons, we find there is a balance between incoming shortwave radiation and outgoing longwave radiation. This is known as radiative equilibrium. The temperature at which the rate of absorption of solar radiation equals the rate of emission of longwave radiation is called the **radiative equilibrium temperature.**

If there were no atmosphere, the radiative equilibrium temperature for the entire Earth would be about 60°F colder than it is now with our atmosphere. The difference is due to the absorption of infrared radiation by greenhouse gases and the fact that these gases radiate infrared energy back to Earth. Without the atmospheric greenhouse effect, Earth could not maintain a temperature that supports life as we know it.

C. WATER VAPOR AND TEMPERATURE

The principle of the atmospheric greenhouse effect can be applied to a particular location. The absorption and emission of infrared radiation by water vapor in the atmosphere is an important factor in predicting how much the temperature will fall at night.

For question 13 below, consider two cities, X and Y. At city X, the air contains very little water vapor. At city Y, the air contains a high amount of water vapor. Based on this information, assume you have to forecast the minimum temperature for tonight at both cities.

13. Other factors being equal, will the minimum temperature tonight at city X be higher or lower than at city Y? Briefly give a reason for your answer.

Name: _____

Course Number _____ Section Number _____

ANSWER SHEET FOR LAB 3

1. At what latitude is the solar angle 90° on the Northern Hemisphere winter solstice?

2. Assume a city is located at a latitude of 35°N. During the year, will the solar angle at that city ever be 90°? If you answer yes, during what month would a solar angle of 90° occur?

3. Compute the solar angle at International Falls, Minnesota (INL) and Brownsville, Texas (BRO) for the dates given below. The latitude of International Falls is 48.6°N and the latitude of Brownsville is 25.9°N.

	BRO	INL
June 20/21	_____	_____
December 21/22	_____	_____

4. On the summer solstice, which location (BRO or INL) has the higher solar angle?

5. On the winter solstice, which location (BRO or INL) has the higher solar angle?

6. On both the summer and winter solstice, which location (BRO or INL) is closer to the latitude where the solar angle is 90º?

7. Assume it's the winter solstice and you are at the latitude where the solar angle is 90º. If you could compute the solar angle at each latitude as you move north to INL, would the solar angles increase or decrease?

8. Which diagram (A, B, or C) best represents the Sun's position in the sky at solar noon at BRO on the summer solstice?

9. Which diagram (A, B, or C) best represents the Sun's position in the sky at solar noon at INL on the winter solstice?

10. At BRO, as you move forward in time from the winter solstice to the summer solstice:
    a. Does the solar angle increase or decrease?

    b. Does the amount of insolation increase or decrease?

11. On the winter solstice:

a. Which location, BRO or INL, receives the lesser amount of insolation?

b. Which location will have more surface heating by the Sun?

12. At point B in Figures 7a and 7b, compare the average temperatures in January and July. The reason why the average temperature at point B in July is higher than the average temperature in January is largely because the solar angle at point B in July is _____ than it is in January. (fill in the blank with either *higher* or *lower*)

13. Other factors being equal, will the minimum temperature tonight at city X be higher or lower than at city Y? Briefly give a reason for your answer.

LAB 4

THE ATMOSPHERE IN MOTION

I. INTRODUCTION

A. OVERVIEW

The atmosphere is a dynamic 3-dimensional fluid with both horizontal and vertical motions. The horizontal motion of the atmosphere, commonly called wind, is a result of forces acting on the air. These forces are responsible for winds at the surface as well as above the surface.

The atmosphere also has vertical motions. Air can rise or sink. This has a great influence on cloud formation, as will be explained in Lab 5.

In this lab, we will look at three forces governing horizontal motions in the atmosphere. Important points regarding each force will be presented, followed by a discussion of how the combination of forces produces a resultant wind. Lastly, some applications will be given.

B. OBJECTIVES

Upon completion of this lab, you should be able to:
1. Describe the forces acting on air at the surface and above the surface.
2. Describe the wind flow in relation to isobars both at the surface and above the surface.
3. Describe the vertical movement of air with both high pressure centers and low pressure centers at the surface.

II. SUMMARY OF FORCES

The three forces we will consider are: pressure gradient force, Coriolis force, and frictional force. Each force acts on the air with a certain magnitude (strength) and in a specific direction. The magnitude of each force can vary depending on the situation.

The intent of this section is not to present a detailed explanation of each force. Rather, a brief description of each force will be given to provide enough information to understand how forces combine to produce a resultant wind.

A. PRESSURE GRADIENT FORCE

A change or difference in pressure between two locations is called a **pressure gradient**. Pressure gradients occur at the surface and at altitudes above the surface.

The existence of a pressure gradient results in a force that is directed from high to low pressure. This force is called the pressure gradient force. Isobars on a weather map graphically show the pressure gradient. In Figure 1, the solid lines are isobars. Lower pressure is to the north.

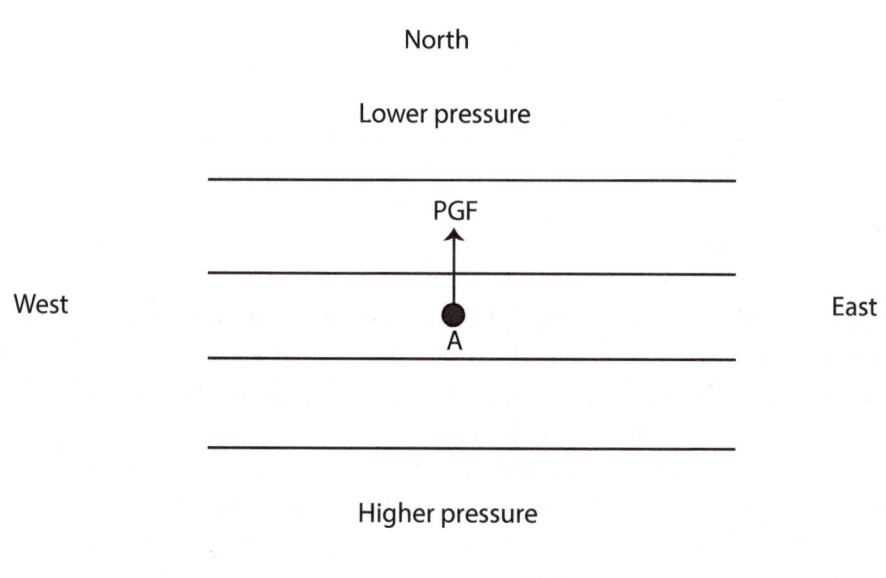

The arrow shows the direction of the pressure gradient force (PGF) at point A. Notice the arrow is drawn perpendicular (at a 90° angle) to the isobars from high to low pressure. There is no pressure gradient parallel to the isobars.

The magnitude of the pressure gradient force is proportional to the difference in pressure over some distance. The greater the pressure difference, the stronger the pressure gradient force. In areas where isobars are closely spaced, the pressure gradient is large and therefore the pressure gradient force is strong because the pressure is changing rapidly with distance. If isobars are farther apart, the pressure gradient force is weaker.

Use Figures 2a and 2b on the answer sheet to answer questions 1 and 2. In both figures, isobars have been drawn for sea level pressures similar to what you might see on a surface weather map. The isobars are drawn using 4 millibar intervals.

1. On each diagram, draw an arrow from point X showing the direction of the pressure gradient force. Label the arrow as "PGF."

2. The magnitude of the pressure gradient force at point X is stronger in which figure? Briefly give a reason for your answer.

In summary, the pressure gradient force initially sets air in motion. It is directed from high to low pressure perpendicular to the isobars. Its magnitude is proportional to the isobar spacing.

B. CORIOLIS FORCE

After air has been moving for a certain amount of time, a second force called the Coriolis force influences the movement of air. The Coriolis force is due to the rotation of the Earth on its axis. Air moving over the surface of Earth will appear to an observer on Earth to be deflected to the right in the Northern Hemisphere following the direction of motion. The deflection is to the left in the Southern Hemisphere. *For this lab, only the Northern Hemisphere will be considered.*

The Coriolis force is a relatively small force, so air has to move for a lengthy period of time before this force has a significant effect. Coriolis force changes the direction of moving air but not the speed. The force is directed at a 90° angle to the right of the wind following the motion. In other words, Coriolis force points at a 90° angle to the wind direction. The magnitude of the Coriolis force depends on wind speed and latitude. The faster air is moving, the stronger is the Coriolis force at a given latitude. The Coriolis force is zero at the equator and increases northward reaching a maximum value at the North Pole. Emphasis in this lab will be on the direction of the Coriolis force.

Below are two examples of how the Coriolis force is directed for the given wind direction. We will use "COR" to abbreviate Coriolis force.

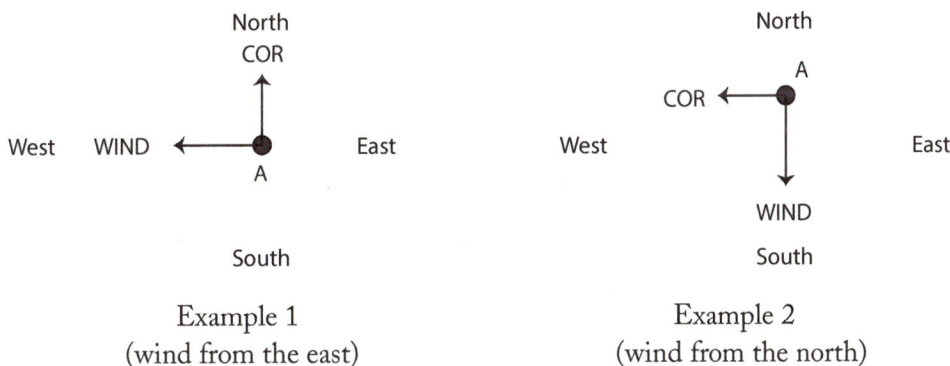

Example 1
(wind from the east)

Example 2
(wind from the north)

In example 2 above, it might appear that the Coriolis force is directed to the left of the wind. However, if you imagine standing at point A facing south with your back to the wind, the Coriolis force is acting to your right or toward the west. Again, Coriolis force acts to the right of the wind, at a 90° angle as you're facing the way the wind is going.

Use Figures 3a and 3b on the answer sheet to answer question 3. In both figures, the wind direction is noted by the arrow drawn from point A.

3. In each figure, draw an arrow showing the direction of the Coriolis force at point A. Label the arrow as "COR."

In summary, the Coriolis force is due to the Earth's rotation. It is directed to the right of the wind following the motion at a 90° angle. Its magnitude depends on wind speed and latitude.

C. FRICTIONAL FORCE

Frictional force, like Coriolis force, doesn't act on air until air starts moving. This force is important for winds at the Earth's surface. The surface as well as other objects act to slow down the wind. At some point above the surface (e.g., 500 meters), the frictional force is negligible and does not need to be considered.

Friction acts to slow the wind so this force is directed opposite to the wind. We will not be concerned with the magnitude of the frictional force in this lab.

III. COMBINING THE FORCES

In this section, we will discuss the combination of forces and resultant wind as they relate to different isobar patterns. We will begin by looking at winds above the surface where the frictional force can be neglected. Then we'll discuss winds at the surface where all three forces must be considered.

A. WINDS ABOVE THE SURFACE

In the atmosphere where the frictional force can be neglected, the wind at some location is governed by the combination of pressure gradient force and Coriolis force. Meteorologists use constant pressure maps to depict weather variables above the surface. On these maps, contour lines are drawn instead of isobars. For this lab, we will use maps showing isobars at a particular altitude since you should already be familiar with isobars.

In the discussion that follows, a few simplifications have been made that should not detract from the intended purpose of this lab. The purpose is to provide a general explanation of why the winds in the atmosphere behave the way they do.

In the diagrams used in this section, PGF stands for pressure gradient force and COR for Coriolis force.

The following three points summarize wind flow above the surface where we can neglect the frictional force:

A. The pressure gradient force and Coriolis force are in balance. This means the two forces act in opposite directions but are equal in magnitude.

B. The wind direction is parallel to the isobars with low pressure to the left.

C. The wind speed is related to the isobar spacing. In general, closely spaced isobars imply faster wind speeds. The pressure gradient force is stronger when isobars are close together.

Consider the following two examples:

In Example 1 below, isobars are drawn as straight lines with lower pressure to the north.

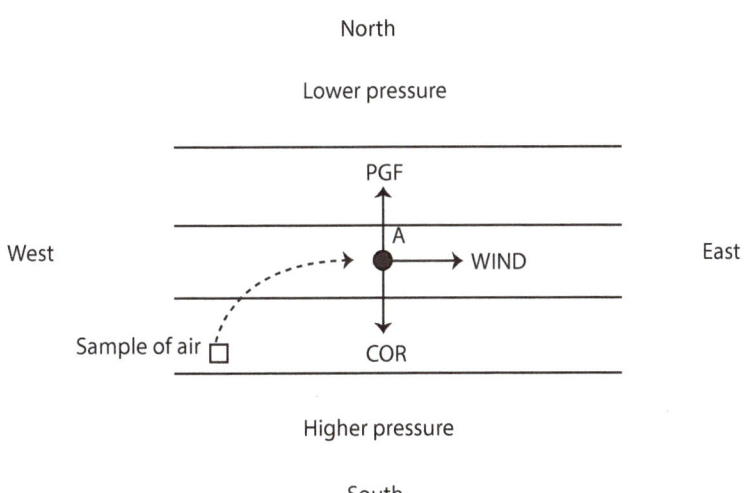

In Example 1, imagine placing a small sample of air (as shown by the small square) into this pressure gradient system. It begins accelerating northward due to the pressure gradient force acting toward lower pressure. As the air gains speed, the Coriolis force increases since Coriolis force depends on wind speed. This causes the sample of air to be deflected to the right following the motion. The dashed line traces the path of the sample of air. Eventually the sample of air reaches point A where the two forces are in balance and the air continues moving parallel to the isobars.

Example 2 below shows curved isobars depicting a low pressure center and a high pressure center. Low pressure and high pressure centers were discussed in Lab 1.

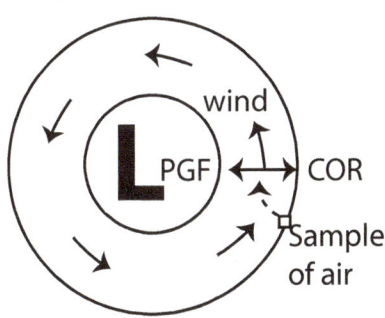

Low pressure center

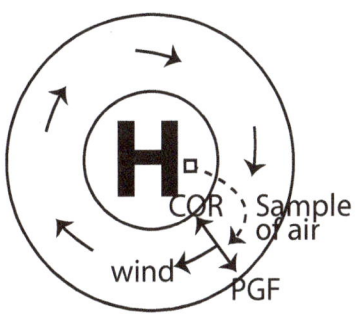

High pressure center

First, consider the low pressure center on the left. If we hypothetically place a sample of air into this pressure gradient system and follow its motion, we would see that the air initially accelerates toward the center of low pressure due to the pressure gradient force. Coriolis force deflects the air to the right. Eventually the air flows parallel to the isobars around the low pressure center in a counterclockwise sense.[1]

For the high pressure center depicted on the right, a sample of air would initially accelerate away from the center due to the pressure gradient force (high to low pressure). Then the Coriolis force would deflect the air to the right. Eventually the air flows parallel to the isobars around the high pressure center in a clockwise sense.

Use Figure 4 on page 61 to answer questions 4 through 6.

Figure 4 is a map showing isobars representing pressures in the troposphere at 18,000 feet. The arrows drawn on the map represent the wind direction.

4. Which point (A, B, or C) has the highest wind speed?

5. At point B, draw and label two arrows, one representing the pressure gradient force (PGF) and the other representing the Coriolis force (COR).

6. There are four dots labeled 1, 2, 3, and 4 where no wind direction is given. At those points, draw an arrow showing the correct wind direction.

B. WINDS AT THE SURFACE

Surface winds are governed by all three forces: pressure gradient, Coriolis, and frictional. The primary difference between winds at the surface and above the surface is that winds at the surface flow across the isobars rather than parallel to the isobars. The reason is due to the effect of friction. Since the frictional force acts opposite to the wind direction, friction will reduce the wind speed. This will cause the Coriolis force to weaken since the magnitude of the Coriolis force depends on wind speed. The pressure gradient force remains unchanged. So by comparison, pressure gradient force is now stronger than Coriolis force. This means winds will flow across the isobars toward lower pressure. Figure 5 shows the balance of forces for surface winds.

---

1. For sharply curved flow, another force called the centrifugal force is important. However, for purposes of this lab, the centrifugal force will not be considered.

North

Lower pressure

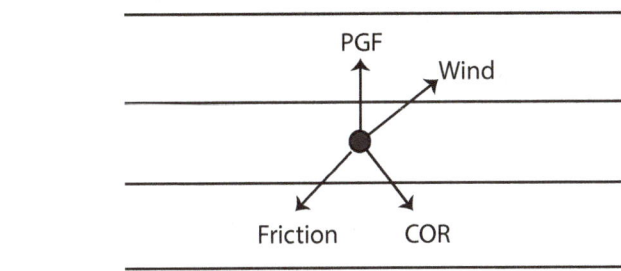

PGF

Wind

West

East

Friction   COR

Higher pressure

South

The three examples below show the direction of surface winds for straight-line isobars and for a high and low pressure center.

North

Lower pressure

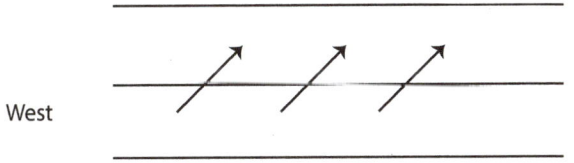

West

East

Higher pressure

South

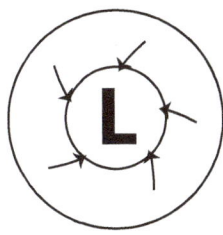 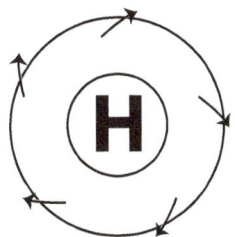

Note that for a low pressure center, the winds flow in a counterclockwise sense inward toward the low center. In the case of a high pressure center, winds flow in a clockwise sense outward away from the high center. This has important implications that will be discussed in the next section.

FIGURE 5

Balance of forces for surface winds. Horizontal lines are isobars.

IV. APPLICATIONS

In this section, we will look at a few applications of the material presented in this lab.

A. SURFACE WINDS AND VERTICAL MOTION

A low pressure center at the surface is characterized by air flowing inward toward the low from different directions. This is called **convergence**. Converging winds at the surface are associated with rising air.

A high pressure center at the surface is characterized by air flowing away from the high in different directions. This is called **divergence**. Diverging winds at the surface are associated with sinking air. Air from above the high pressure center sinks to replace the air flowing horizontally away from the high.

Figure 6 illustrates the coupling of horizontal and vertical motions.

FIGURE 6

Low-level Convergence, Divergence, and Vertical Air Movement

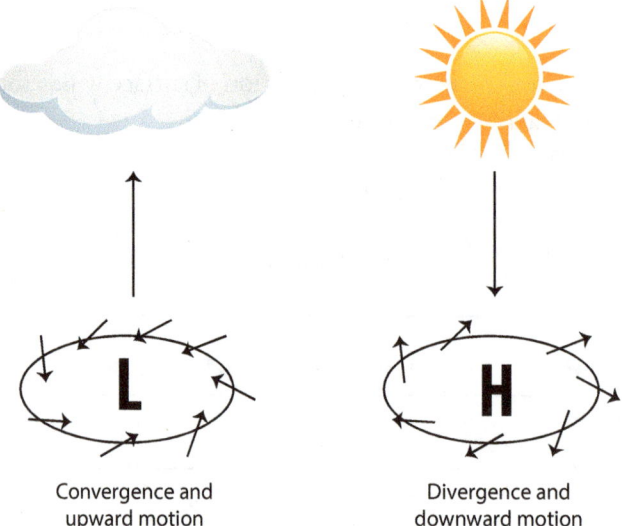

Convergence and upward motion

Divergence and downward motion

With a few exceptions, vertical wind speeds are much weaker than horizontal wind speeds that are routinely measured.

Rising air can lead to the formation of clouds. Sinking air is not favorable for cloud formation. An explanation of how clouds form in the atmosphere will be given in Lab 5.

B. FORECASTING WIND DIRECTION

High and low pressure centers typically move across the U.S. from west to east. The circulation around these high and low pressure centers largely dictates what the wind direction is at a given location. The wind direction will change depending on where a given location is with respect to either a high or low pressure center.

Figure 7 is a surface weather map with isobars drawn to sea level pressures. Assume this map depicts the current conditions. In questions 7 and 8, you will be determining the wind direction at point A in north Texas.

7. Based on the isobar pattern shown in Figure 7, the most likely wind direction at point A is from the
   A. North
   B. South
   C. East
   D. Northeast

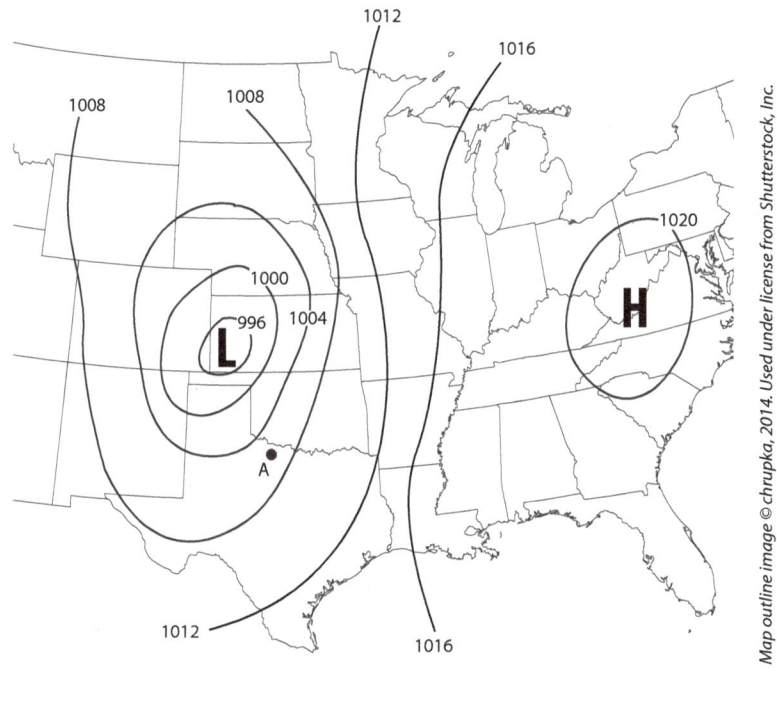

Map outline image © chrupka, 2014. Used under license from Shutterstock, Inc.

FIGURE 7

Current surface weather map. Solid lines are isobars.

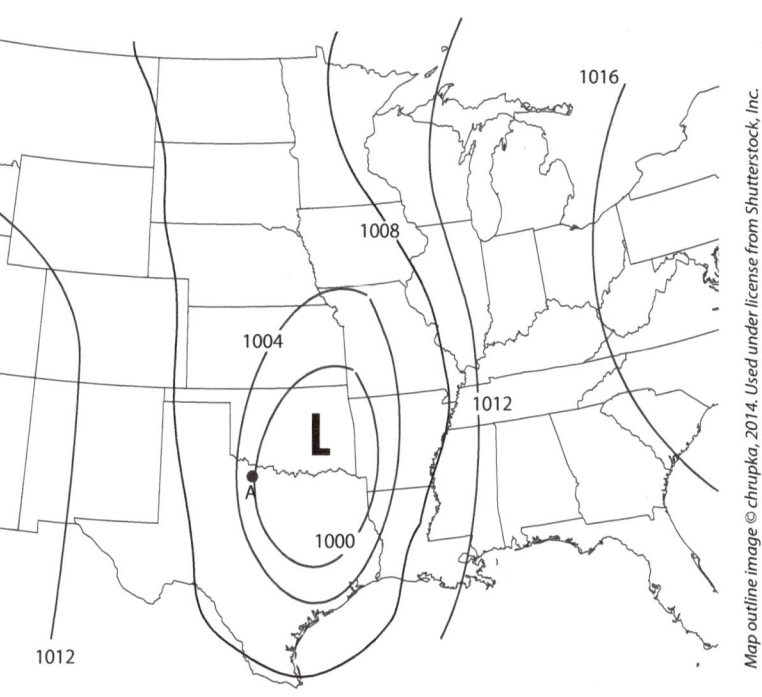

Map outline image © chrupka, 2014. Used under license from Shutterstock, Inc.

FIGURE 8

Surface weather map for 24 hours after Figure 7.

Weather forecasters use computer models of the atmosphere to help predict changes in wind among other weather variables. Figure 8 represents a surface weather map produced by a computer showing isobars for 24 hours after Figure 7.

8. Based on this computer-generated map, what wind direction would you forecast for point A for tomorrow?
    A. South
    B. West
    C. Southeast
    D. Northwest

C. FORECASTING THE MOVEMENT OF HURRICANES

A hurricane is an intense low pressure system with sustained winds of at least 74 mph. They form over warm tropical water where the water temperature is typically 80°F or higher. The movement of hurricanes is partially determined by the larger scale winds. Larger scale refers to the global (planetary) scale.

Over the North Atlantic Ocean there exists a semi-permanent area of high pressure that is part of the subtropical high pressure zone around the globe. An illustration is shown in Figure 9 on page 63.

Hurricanes are much smaller in size compared to this large semi-permanent high pressure system. The winds circulating around the high will help steer the hurricane in a certain direction.

9. The solid lines in Figure 9 represent isobars at some altitude above the surface. At the two dots, draw an arrow representing the wind direction.
10. The symbol for a hurricane is ⑂. In Figure 9, note the position of the hurricane. If the isobar pattern did not change over the next day or so, what direction would you forecast the hurricane to move?

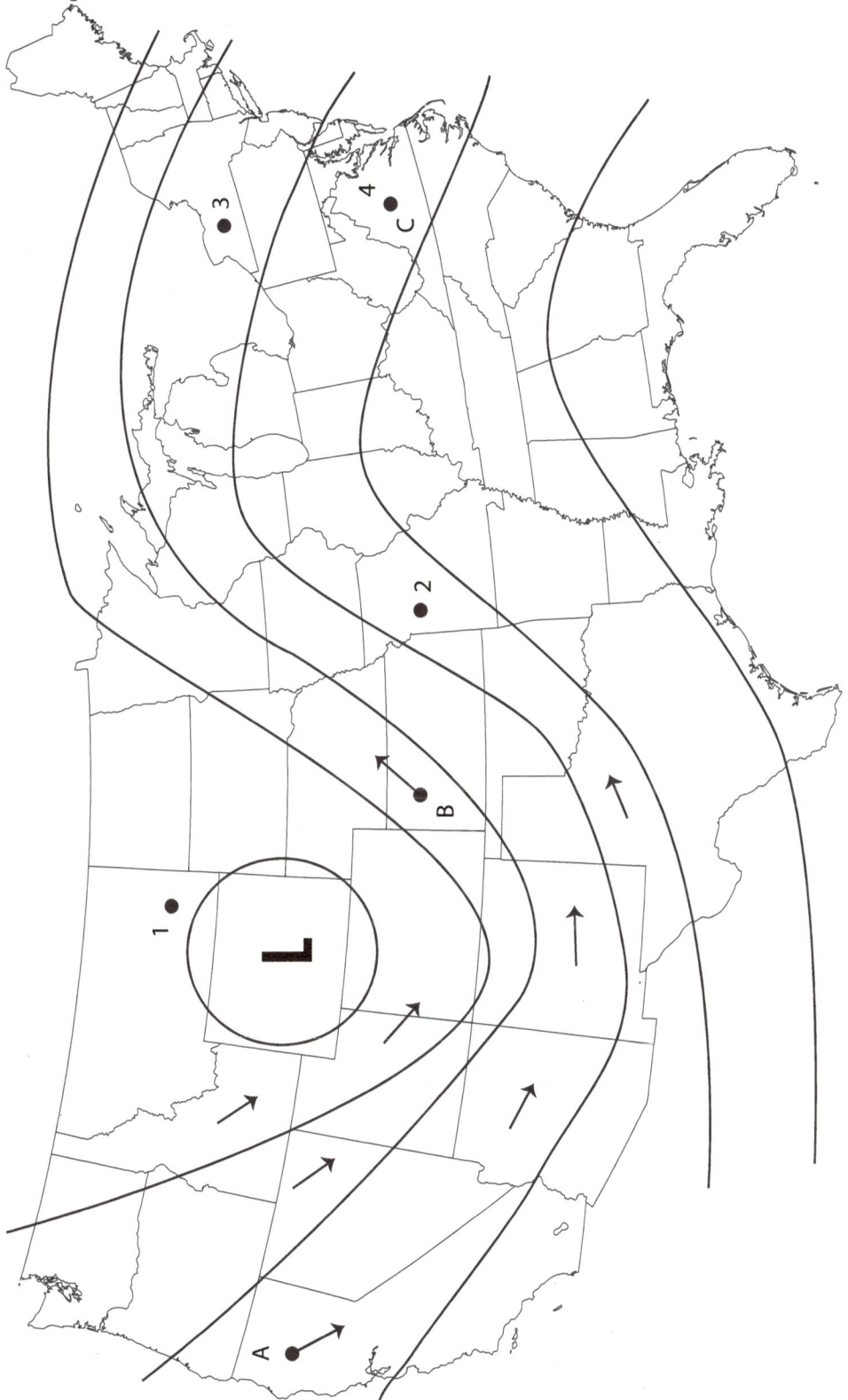

FIGURE 4

Isobars (solid lines) at 18,000 feet. Arrows represent wind direction.

Map outline image © chrupka, 2014. Used under license from Shutterstock, Inc.

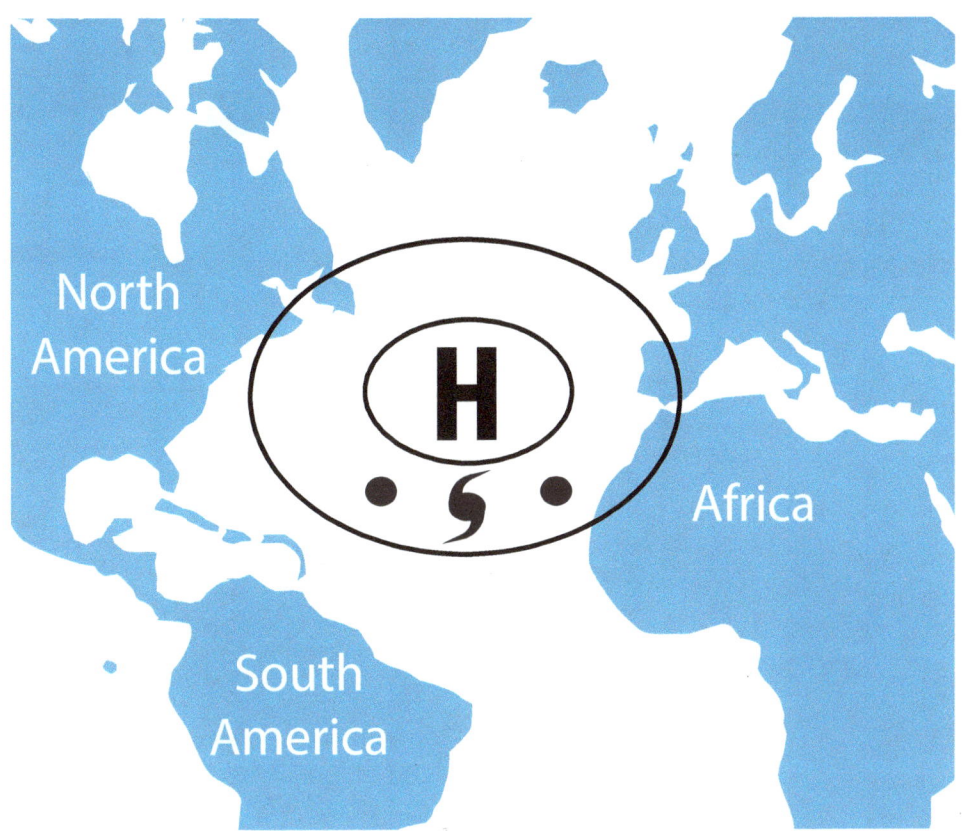

FIGURE 9

Subtropical High Pressure Over the North Atlantic

Name: _____

Course Number _____ Section Number _____

ANSWER SHEET FOR LAB 4

Use Figures 2a and 2b below to answer questions 1 and 2.

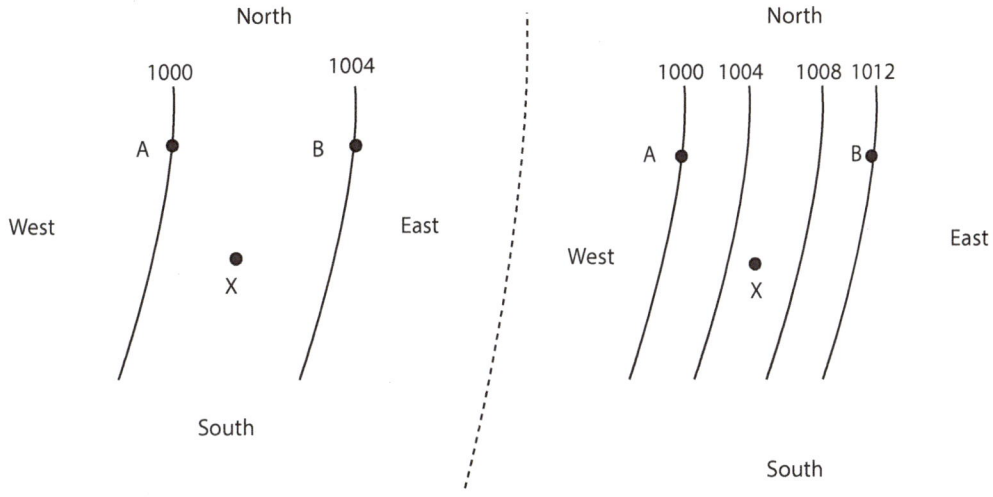

1. On each diagram, draw an arrow from point X showing the direction of the pressure gradient force. Label the arrow as "PGF."

2. The magnitude of the pressure gradient force at point X is stronger in which figure? Briefly give a reason for your answer.

Use Figures 3a and 3b below to answer question 3. In both figures, the wind direction is noted by the arrow drawn from point A.

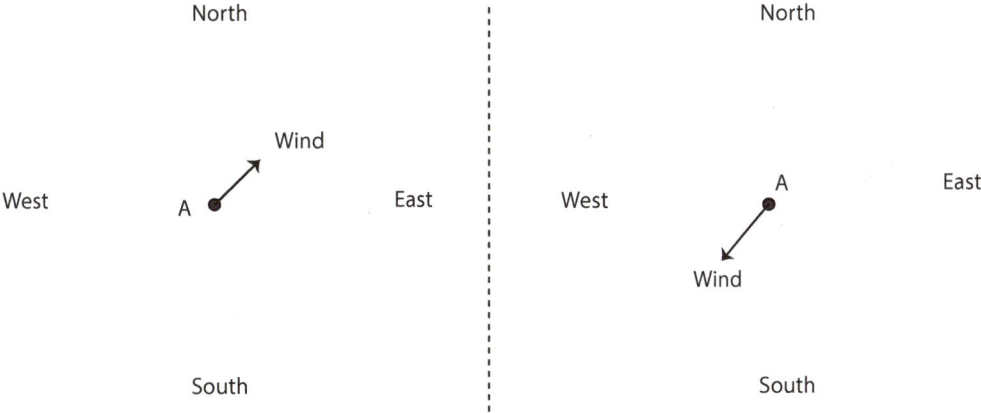

3. In each figure, draw an arrow showing the direction of the Coriolis force at point A. Label the arrow as "COR."

Use Figure 4 on page 61 to answer questions 4 through 6.

4. Which point (A, B, or C) has the highest wind speed?

5. At point B, draw two arrows, one representing the pressure gradient force and the other representing the Coriolis force.

6. There are four dots labeled 1, 2, 3, and 4 where no wind direction is given. At those points, draw an arrow showing the correct wind direction.

7. Based on the isobar pattern in Figure 7, the most likely wind direction at point A is from the
    A. North
    B. South
    C. East
    D. Northeast

8. Based on the computer-generated map (Figure 8), what wind direction would you forecast at point A for tomorrow?
    A. South
    B. West
    C. Southeast
    D. Northwest

9. The solid lines in Figure 9 on page 63 represent isobars at some altitude above the surface. At the two dots, draw an arrow representing the wind direction.

10. The symbol for a hurricane is ᕙ. In Figure 9, note the position of the hurricane. If the isobar pattern did not change over the next day or so, what direction would you forecast the hurricane to move?

LAB 5

HUMIDITY AND CLOUD FORMATION

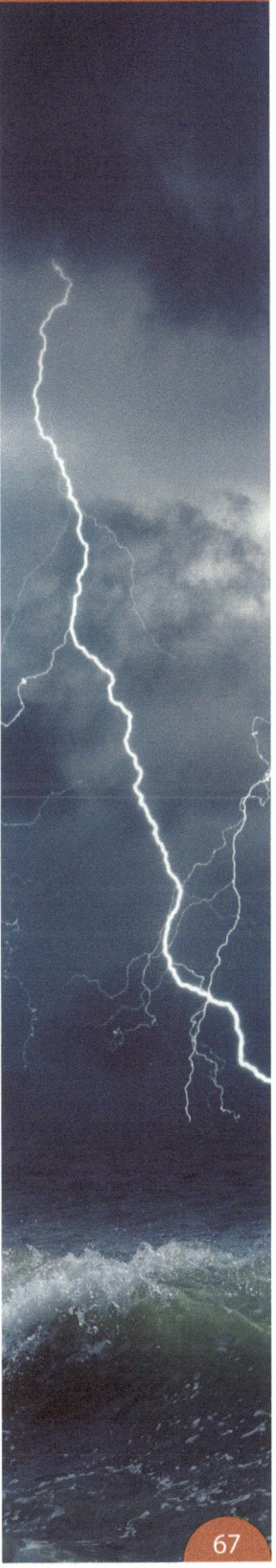

I. INTRODUCTION

A. OVERVIEW

Humidity is a general term that refers to the amount of water vapor in the air. Water vapor is one of the invisible gases that make up the composition of our atmosphere. Some people may refer to the cloud of steam that rises from a pan of boiling water as water vapor, but that is incorrect. You cannot see water vapor molecules.

Water (H_2O) can exist in three states (or phases) in the atmosphere. These are: solid (ice), liquid (usually just called water), and gas (water vapor). Water can change from one phase to another. **Evaporation**, the phase change from liquid to gas, and **condensation**, from gas to liquid, are two examples of a phase change. Phase changes of water in the atmosphere are extremely important. Evaporation adds water vapor to the air. Condensation is important in the formation of clouds.

This lab begins with a description of four terms related to water vapor. These terms will be used to explain how clouds form.

B. OBJECTIVES

Upon completion of this lab, you should be able to:
1. Define terms used in discussing water vapor and explain how each term is used.
2. Define saturation and explain how saturation is achieved in the atmosphere in order for clouds to form.
3. Define and explain the dry and saturated adiabatic lapse rates.

II. SATURATION

The term **saturation** will be used several times in this lab. The important point to remember regarding saturation is that saturated air contains the maximum possible amount of water vapor for a given temperature.

III. TERMS RELATED TO WATER VAPOR

The terms used in this lab to discuss water vapor are mixing ratio, saturation mixing ratio, relative humidity, and dew point temperature.

A. MIXING RATIO AND SATURATION MIXING RATIO

Mixing ratio is a measure of the amount of water vapor in the air. It is defined as follows:

$$\text{Mixing Ratio} = \frac{\text{mass of water vapor}}{\text{mass of dry air}}$$

The term dry air in the above equation refers to all of the gases in the atmosphere except water vapor. The units of mixing ratio are usually given as grams per kilogram (g/kg). When referring to the actual amount (mass) of water vapor in the air, the term **actual mixing ratio** will be used.

The **saturation mixing ratio** is the mass of water vapor to the mass of dry air when the air is saturated. It expresses the maximum possible amount of water vapor that could be in the air at a specified temperature.

It is important to realize that even though mixing ratio (or actual mixing ratio) and saturation mixing ratio are both calculated using the above equation, they are two separate quantities.

Table 1 shows the value of saturation mixing ratio for selected temperatures in degrees Celsius. The atmospheric pressure is assumed to be 1000 mb.

TABLE 1

Saturation Mixing Ratio (g/kg) as a Function of Temperature (°C)

Temperature (°C)	Saturation Mixing Ratio (g/kg)
-10	1.8
-5	2.6
0	3.8
5	5.5
10	7.7
15	10.8
20	14.9
25	20.3
30	27.5
35	36.9
40	49.4

Use Table 1 to answer the following questions:
1. Consider a sample of air with a temperature of 20°C. The air is not saturated.
    a. What is the maximum possible amount of water vapor that could be in this sample of air?

    b. Does the sample of air at 20°C actually contain that amount of water vapor? Briefly explain your answer.

B. RELATIVE HUMIDITY

Relative humidity is a ratio of the actual water vapor content to the maximum possible water vapor content for a given temperature. Using the terms already discussed, relative humidity (RH) is calculated using the following equation:

$$RH = \frac{\text{actual mixing ratio}}{\text{saturation mixing ratio}} \times 100\%$$

The value obtained by doing the division is multiplied by 100 to give the relative humidity in percent.

Note that relative humidity depends on two variables: 1) the actual amount of water vapor present in the air (actual mixing ratio) and 2) the maximum possible amount of water vapor that could be in the air (saturation mixing ratio). The actual mixing ratio will change if water vapor is either added to or removed from the air. Saturation mixing ratio will change as the air temperature changes, as evident from Table 1. Relative humidity will change as air temperature changes even when the actual mixing ratio remains constant. Therefore, relative humidity should *not* be used as a measure of actual water vapor content.

Relative humidity is used to gauge how close the air is to being saturated. When air is saturated, the air contains its maximum possible amount of water vapor. In other words, when air is saturated, the actual mixing ratio equals the saturation mixing ratio and the relative humidity is 100%. If there is a large difference between actual mixing ratio and saturation mixing ratio, the relative humidity will be much lower than 100%.

When saturation occurs at the surface of the Earth, dew, frost, or fog may form. Above the surface, saturation results in the formation of clouds.

C. DEW POINT TEMPERATURE

Dew point temperature, often referred to as just *dew point*, is the temperature to which air must be cooled at constant pressure to achieve saturation. It represents what the air temperature must be for the air to be saturated. If air is cooled to the dew point by some process, then temperature will equal dew point and the air is saturated. Dew point will either be equal to or less than the air temperature.

Use Table 1 and the equation for RH to answer the following questions:

2. Assume the air temperature is 25°C and the RH is 53%.
    a. What is the actual mixing ratio?

Since the RH is just above 50%, the air contains approximately half the amount of water vapor it could contain for a temperature of 25°C.

    b. Let's say the air temperature cools from 25°C to 15°C without any change to the actual water vapor content in the air. At 15°C, what is the saturation mixing ratio?

    c. Using your answers to a and b above, what is the relative humidity at 15°C?

    d. In this situation, the temperature of 15°C is called what?

Table 1 can be used to determine the amount of water vapor in the air (the actual mixing ratio) if the dew point is known. In Table 1, simply let the values in the first column represent dew point temperature and the values in the second column represent actual mixing ratio. Table 1 can be used for two purposes: either to determine the saturation mixing ratio if given a temperature or to determine the actual mixing ratio if given a dew point.

Since dew point is associated with actual mixing ratio, we can say that dew point is an indicator of the actual water vapor content. The higher the dew point, the more water vapor is in the air. This is important to remember.

To help reinforce the relationship among temperature, dew point, and relative humidity, consider Table 2. This table shows hourly surface observations of temperature, dew point, and relative humidity at one location on a particular day. Temperature and dew point values are in degrees Fahrenheit. Use the data in this table to answer questions 3 through 5.

TABLE 2

Hourly Surface Observations of Temperature (°F), Dew Point (°F), and RH (%)

TIME	TEMP	DEW POINT	RH
2:00 am	75	62	64
3:00 am	72	62	73
4:00 am	69	62	81
5:00 am	67	62	84
6:00 am	66	62	87
7:00 am	65	62	90
8:00 am	70	64	81
9:00 am	73	65	76
10:00 am	79	65	62
11:00 am	83	62	49

TABLE 2

Continued...

TIME	TEMP	DEW POINT	RH
Noon	86	61	43
1:00 pm	91	59	34
2:00 pm	95	56	27
3:00 pm	96	54	24
4:00 pm	97	45	17
5:00 pm	97	45	17
6:00 pm	95	46	19
7:00 pm	95	46	19
8:00 pm	95	46	19
9:00 pm	92	47	21
10:00 pm	78	55	45
11:00 pm	72	60	66

3. Between 2:00 am and 7:00 am, the dew point did not change. What does this indicate regarding the amount of water vapor in the air?

4. Why did the relative humidity increase between 2:00 am and 7:00 am?

Changes in relative humidity do not always imply a change in the amount of water vapor in the air.

5. From 9 pm to 11 pm, did the water vapor content of the air increase or decrease? Explain your answer.

IV. CLOUD FORMATION

A. CONDENSATION

Clouds can form at different altitudes in the troposphere. They can consist of tiny ice crystals, tiny liquid water drops, or both. For this lab, we will focus on clouds composed of liquid water. The liquid water drops form from condensation (gas to liquid). For condensation to occur, the air must be saturated, meaning the relative humidity is 100%. Therefore, cloud formation requires unsaturated air to become saturated.

B. GETTING THE AIR SATURATED

The primary way of achieving saturation so clouds can form is cooling the air. This cooling occurs as air rises. To explain how rising air results in cooling, imagine a representative sample of air about the size of a large balloon or bubble. This is called a **parcel of air**. The idea of a parcel of air is a useful tool to explain atmospheric processes. As the air parcel moves through the atmosphere, there is no exchange of heat or mass with its surroundings (environment).

A parcel of air contains a certain amount of water vapor and has a certain temperature. When an air parcel rises, it expands. It expands because as the parcel rises, the pressure around it decreases. The expansion uses energy that causes the temperature in the parcel to drop.

The cooling that results from the parcel expanding as it rises is called an **adiabatic** process. Adiabatic means there is no heat energy exchange between the parcel and its surrounding environment. The parcel cools because it is expanding.

There are several mechanisms that can cause air to rise. One was discussed in Lab 4. As winds spiral inward (convergence) around a low pressure center at the surface, air rises. Another method of forcing air to rise occurs when air is moving toward higher terrain, such as the Rocky Mountains. Air is forced upward by a mountain as seen in Figure 1 below. This is called *upslope flow*.

FIGURE 1

Side view of a mountain with upslope flow. Arrows represent winds.

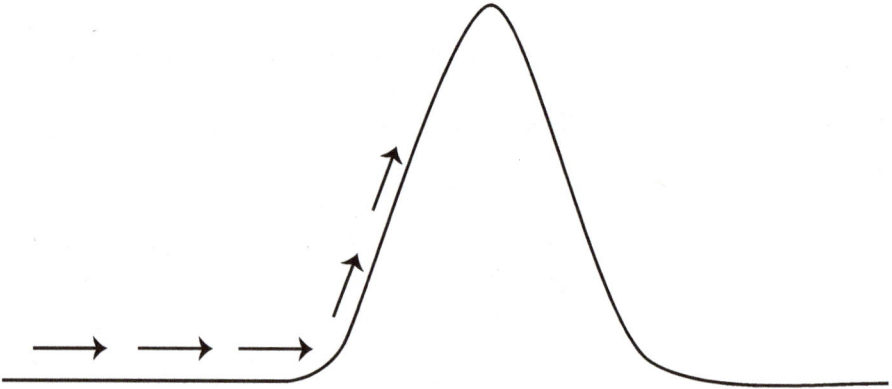

Conversely, sinking air parcels will warm adiabatically. When an air parcel sinks in the atmosphere, it is compressed. That results in an increase in the parcel temperature.

In Lab 4, it was explained that downward motion is associated with high pressure centers at the surface. The adiabatic warming that occurs with sinking air does not lead to saturation and cloud formation. That is why high pressure centers are generally characterized by an absence of extensive clouds and precipitation.

Consider a parcel of air at the surface that is not saturated (unsaturated). If it is forced to rise, its temperature will decrease at a constant rate. The rate of cooling is called the **dry adiabatic lapse rate**. In this term, "dry" implies the air is not saturated. The dry adiabatic lapse rate is equal to 1°C/100 meters or 10°C/km.

To illustrate the dry adiabatic lapse rate, we will look at a parcel of air originally at the surface. Figure 2 shows an unsaturated parcel of air at the surface with a temperature of 86°F (30°C).

FIGURE 2

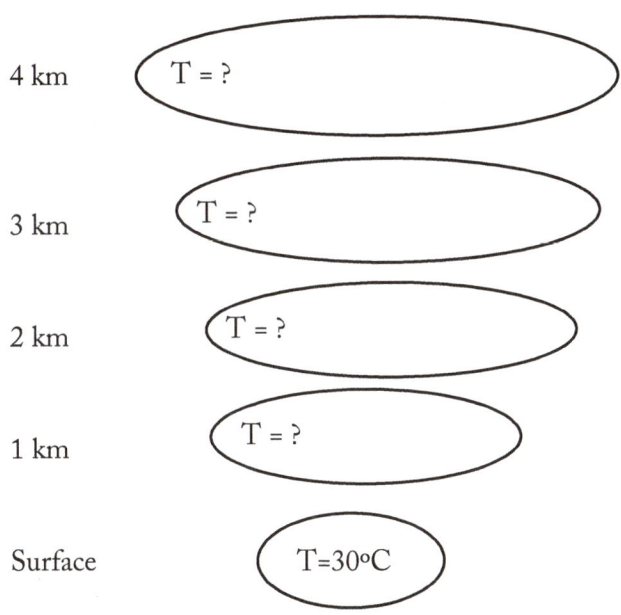

4 km   T = ?

3 km   T = ?

2 km   T = ?

1 km   T = ?

Surface   T=30ºC

6. If the parcel at the surface in Figure 2 were forced to rise, what would its temperature be at 1, 2, 3, and 4 kilometers? (The parcel remains unsaturated throughout the ascent.)

We will use Figure 3 to illustrate the cloud formation process. You will be asked to fill in values for several moisture variables in a rising parcel of air. Note that the figure includes a surface dew point (Td) to account for the amount of water vapor in the air. Dew point is an indicator of the actual water vapor content. Table 1 on page 68 can be used to determine the actual mixing ratio (right column) when given the dew point (left column) as explained earlier. In Figure 3, AMR stands for actual mixing ratio, SMR for saturation mixing ratio, and RH for relative humidity.

Questions 7 through 12 pertain to Figure 3.

FIGURE 3

Rising Parcel of Air

4 km   T=_____   Td=_____   AMR=_____   SMR=_____   RH=_____

3 km   T=_____   Td=_____   AMR=_____   SMR=_____   RH=_____

2 km   T=_____   Td=_____   AMR=_____   SMR=_____   RH=_____

1 km   T=_____   Td=_____   AMR=_____   SMR=_____   RH=_____

Surface   T=30   Td=10   AMR=_____   SMR=_____   RH=_____

7. The surface dew point is 10°C (50°F). Using Table 1, what is the actual mixing ratio (AMR) and the saturation mixing ratio (SMR) at the surface? Write your answer in the figure on the answer sheet.

8. Compute the relative humidity at the surface and write your answer in the figure on the answer sheet.

9. Is air at the surface saturated or unsaturated?

Now consider the parcel of air rising from the surface. In a rising parcel of unsaturated air, the amount of water vapor does not change. This means that the actual mixing ratio does not change. The dew point does change slightly due to changes in pressure, but for this lab we will assume the dew point remains constant in rising, unsaturated air. The saturation mixing ratio will change because the temperature changes.

10. In the figure on the answer sheet, write in the temperature, dew point, actual mixing ratio, saturation mixing ratio, and relative humidity in the parcel at 1 km and 2 km.

11. At 2 km, the parcel becomes saturated. From Figure 3, what indicates the parcel is saturated?

The altitude where a rising parcel of unsaturated air first becomes saturated is called the **Lifting Condensation Level (LCL)**. At this level, the temperature of the parcel has cooled to the dew point. The air is saturated so condensation begins and the cloud base starts forming. The water vapor molecules in the air start to condense onto tiny particles called condensation nuclei.

If the saturated parcel continues to rise, the parcel will continue to expand and cool. However, the rate of cooling will be less than the dry adiabatic lapse rate. This is because as condensation occurs, latent heat energy is released in the parcel, which slows or offsets the cooling due to expansion. The rate of cooling of rising saturated air is called the **saturated adiabatic lapse rate**. It is not a constant, but varies depending on the amount of water vapor. For this lab, a value of 5°C/km will be used.

The relative humidity in the parcel as it rises above the LCL remains at 100%. This is the relative humidity in the cloud. With this information, answer question 12.

12. In the figure on the answer sheet, fill in the temperature, dew point, AMR, SMR, and RH at 3 km and 4 km.

As the parcel rises above the LCL, the actual mixing ratio decreases. This is because some of the water vapor molecules are being converted to liquid water drops by the condensation process.

We have shown how clouds form when parcels of air at the surface are lifted in the atmosphere. Clouds can form when air is lifted from an altitude somewhere above the surface as well.

You should now have a basic understanding of the process leading to cloud formation in the atmosphere. In Lab 6, we will discuss different types of clouds and the controlling factor that determines the type of cloud that forms.

Name: _____

Course Number _____ Section Number _____

ANSWER SHEET FOR LAB 5

1. Consider a sample of air with a temperature of 20°C. The air is not saturated.
   a. What is the maximum possible amount of water vapor that could be in this sample of air (use Table 1)?

   b. Does the sample of air at 20°C actually contain that amount of water vapor? Briefly explain your answer.

2. Assume the air temperature is 25°C and the RH is 53%.
   a. What is the actual mixing ratio? (round up to the nearest tenth)

   b. Let's say the air temperature cools from 25°C to 15°C without any change to the actual water vapor content in the air. At 15°C, what is the saturation mixing ratio?

c. Using your answers to a and b above, what is the relative humidity at 15°C?

d. In this situation, the temperature of 15°C is called what?

Use Table 2 for questions 3, 4, and 5.

3. Between 2:00 am and 7:00 am, the dew point did not change. What does this indicate regarding the amount of water vapor in the air?

4. Why did the relative humidity increase between 2:00 am and 7:00 am?

5. From 9 pm to 11 pm, did the water vapor content of the air increase or decrease? Explain your answer.

6. In the diagram below (Figure 2 in the text), fill in the temperature of the unsaturated parcel of air at each altitude as it rises from the surface to 4 km.

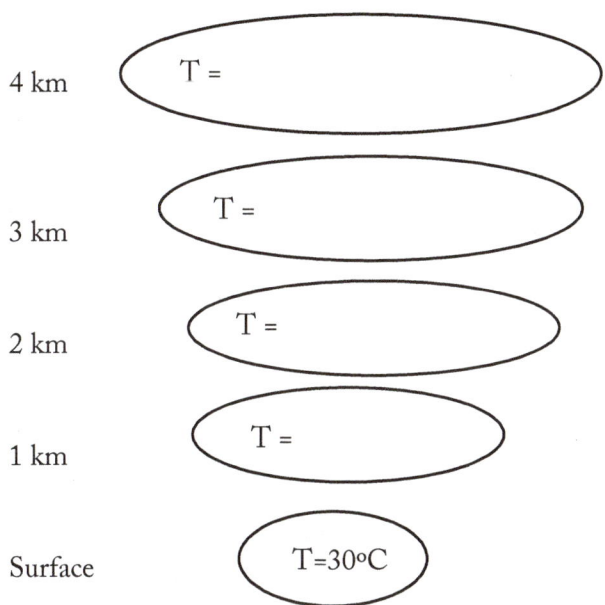

4 km   T =

3 km   T =

2 km   T =

1 km   T =

Surface   T=30°C

Questions 7 through 12 pertain to the figure below (Figure 3 in the text):

4 km   T=_____   Td=_____   AMR=_____   SMR=_____   RH=_____

3 km   T=_____   Td=_____   AMR=_____   SMR=_____   RH=_____

2 km   T=_____   Td=_____   AMR=_____   SMR=_____   RH=_____

1 km   T=_____   Td=_____   AMR=_____   SMR=_____   RH=_____

Surface   T=30   Td=10   AMR=_____   SMR=_____   RH=_____

7. The surface dew point is 10°C (50°F). Using Table 1, what is the actual mixing ratio (AMR) and the saturation mixing ratio (SMR) at the surface? Write your answer in the figure above.

8. Compute the relative humidity at the surface and write your answer in the figure above.

9. Is air at the surface saturated or unsaturated?

10. In the figure above, write in the temperature, dew point, actual mixing ratio, saturation mixing ratio, and relative humidity in the parcel at 1 km and 2 km.

11. At 2 km, the parcel becomes saturated. What indicates the parcel is saturated?

12. In the figure above, fill in the temperature, dew point, AMR, SMR, and RH at 3 km and 4 km.

LAB 6

ATMOSPHERIC STABILITY AND CLOUD TYPES

I. INTRODUCTION

A. OVERVIEW

Clouds have a variety of appearances. Some clouds form individually with well-defined edges and can extend vertically through much or all of the troposphere. Other clouds appear as a layer covering the sky with little vertical development.

The way a cloud develops once air becomes saturated depends on a property of the atmosphere called **stability**. In this lab, we will investigate the important role of atmospheric stability. A few cloud types will also be presented.

B. OBJECTIVES

Upon completion of this lab, you should be able to:
1. Define and state the importance of atmospheric stability.
2. Explain how to determine whether a parcel of air is stable, unstable, or neutral.
3. Identify selected cloud types.

II. ATMOSPHERIC STABILITY

A. DESCRIPTION

The term stability refers to whether or not the atmosphere favors vertical air motion or resists it. Parcels of air can be stable or unstable. They can also be neutral in terms of stability, but that will be defined later. If an unstable parcel is displaced vertically, the displacement is amplified, meaning the parcel accelerates in the direction of the displacement. If a stable parcel is displaced vertically, it will resist that displacement and will tend to return to its original position. So with unstable parcels, vertical motion is favored. With stable parcels, vertical motion is hindered.

Stability is a characteristic of the atmosphere at any given time and place. It may change from one day to the next or even change from morning to afternoon at a particular location.

B. IMPORTANCE

The stability of the atmosphere largely determines the type of cloud that forms when saturated air rises. Clouds that have a greater vertical extent than horizontal extent are associated with unstable parcels of air, often referred to as **instability**. These clouds are generally called *cumuliform* clouds. Figure 1 is an example of a cloud with extensive vertical development. This cloud is called a *cumulus congestus* or towering cumulus.

Image © Condor 36, 2014. Used under license from Shutterstock, Inc.

If the air parcels have limited instability, there won't be as much vertical development to the cloud. In this case, *fair weather cumulus* clouds form (Figure 2).

Image © irin-k, 2014. Used under license from Shutterstock, Inc.

Fair weather cumulus clouds typically form due to surface heating. They are sometimes seen on a summer afternoon as solar radiation heats the ground. Air in contact with the ground is heated by conduction. The air then becomes buoyant and rises. This upward transport of heat energy is called **convection**. Convection is an energy transfer process by the movement of air. If the air contains sufficient moisture and can rise to the level of saturation (LCL), a cloud forms.

Clouds that have a large horizontal extent but not much vertical extent are associated with stable parcels of air. These clouds are generally called *stratiform* clouds. Figure 3 is an example of a *stratus* cloud. Stratus clouds usually appear as a solid layer covering all or most of the sky.

FIGURE   3

Source: Loren Phillips

Figure 4 shows the stratus clouds beginning to dissipate. You can see that the clouds have very little vertical development.

FIGURE   4

Source: Loren Phillips

C. HOW STABILITY IS DETERMINED

Stability is related to the buoyancy of air parcels. To perhaps better visualize how stability is determined, we'll use a communication tower such as the one in Figure 5. Assume the tower is 200 meters tall.

FIGURE 5

Communication Tower

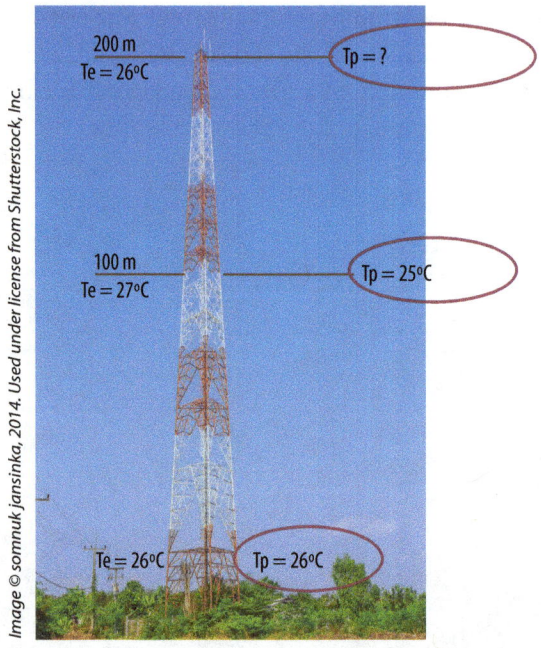

If a thermometer was positioned on the tower at 100 meters and another thermometer at 200 meters, they would show the air temperature. We'll call this temperature the environmental temperature and represent it by the notation Te. Assume the air temperature at the surface is 26°C. The environmental temperature at 100 meters is 27°C and at 200 meters it is 26°C.

Now imagine a parcel of air at the surface with a temperature of 26°C. The parcel temperature is represented by the notation Tp. The parcel is not saturated. If this parcel could be forced up to a level of 100 meters, the parcel temperature would cool at 1°C/100 meters (the dry adiabatic lapse rate). So the parcel temperature at 100 meters becomes 25°C. Compared to the temperature of the environment at 100 meters, the parcel is colder. Cold air is more dense than warm air, so if the mechanism forcing the air to rise were removed, the parcel would sink. The parcel is said to have negative buoyancy and considered to be stable. The tendency of the parcel is to sink, so essentially the atmosphere is resisting vertical motion.

1. In Figure 5, if the parcel were forced from 100 meters up to 200 meters, what would the parcel temperature be at 200 meters?

2. Is the parcel stable or unstable at 200 meters?

Figure 6 is similar to Figure 5, except that the environmental temperature at 100 and 200 meters has changed. The temperature at the surface is still 26°C.

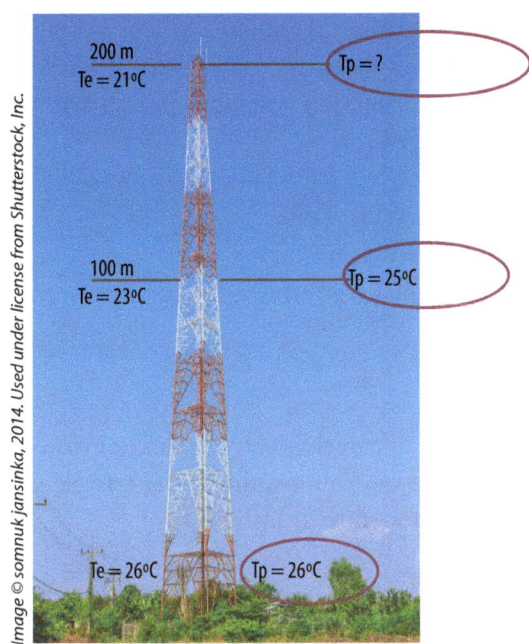

FIGURE 6

Communication
Tower

200 m
Te = 21°C

Tp = ?

100 m
Te = 23°C

Tp = 25°C

Te = 26°C   Tp = 26°C

Image © somnuk jansinka, 2014. Used under license from Shutterstock, Inc.

If an unsaturated parcel at the surface could be lifted to 100 meters, it will again cool at the dry adiabatic lapse rate. So the parcel arrives at 100 meters with a temperature of 25°C. Notice in this case that the parcel temperature is warmer than the surrounding environmental temperature. Since warm air is less dense than cold air, it will have positive buoyancy and continue to rise. The parcel is said to be unstable. It doesn't need a forcing mechanism at this point to lift the air. This is effectively saying the atmosphere promotes rising air. The atmosphere is not resisting rising air, as was the case when the parcel was stable.

3. In Figure 6, as the parcel continues to rise from 100 meters up to 200 meters, what is the parcel temperature at 200 meters?

4. Is the parcel stable or unstable at 200 meters?

There are situations in which the parcel is lifted to a certain altitude and the parcel temperature equals the environmental temperature. In this case, the parcel has a neutral stability.

We have used these simple illustrations to explain how to evaluate the stability of a parcel. In weather forecasting, meteorologists must evaluate the stability throughout the entire troposphere. The radiosonde, described in Lab 2, is used to obtain the environmental temperature at many altitudes through the troposphere into the stratosphere.

It is very important *not* to confuse the parcel temperature with the environmental temperature. The radiosonde measures only the environmental temperature, not the parcel temperature. The parcel temperature is determined using the dry adiabatic lapse rate or the saturated adiabatic lapse rate. In the illustrations using Figures 5 and 6, we kept the parcel unsaturated. If a parcel is saturated, the same principles apply to determine the stability of the parcel.

In summary, to determine whether a parcel is stable, unstable, or neutral at a given altitude, we only need the parcel temperature and the environmental temperature. Using our notation for the parcel temperature (Tp) and the environmental temperature (Te):

if Tp < Te, the parcel is stable
if Tp > Te, the parcel is unstable
if Tp = Te, the parcel is neutral

D. APPLICATION

Evaluating the stability of air parcels is very important in forecasting the development of thunderstorms. Thunderstorms require rising parcels of saturated air that are unstable. Again, this is referred to as instability.

Thunderstorms form when numerous parcels of air at the surface are forced to rise. The parcels are initially stable but soon reach an altitude above which the parcels become unstable. At that point, positive buoyancy allows the parcels to accelerate upward. If a sufficient number of rapidly rising parcels exist, a *cumulonimbus* cloud forms (Figure 7).

FIGURE 7

Cumulonimbus Cloud

Image © swa182, 2014. Used under license from Shutterstock, Inc.

Temperatures obtained from the radiosonde as it is carried upward by a balloon can be plotted on a diagram similar to the one in Figure 8. These are the environmental temperatures.

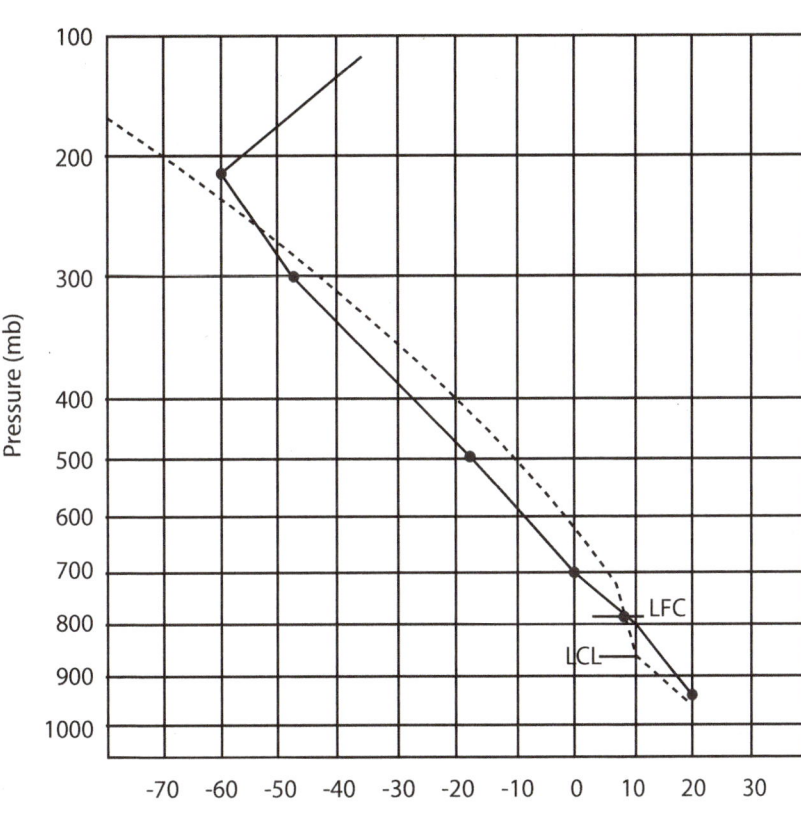

The solid line in Figure 8 shows how the air temperature (environmental temperature) changes with height. In this example, the pressure at the surface is 950 millibars and the surface temperature is 20°C (68°F).

To evaluate stability, the temperature in a rising parcel of air must be known. We will assume air at the surface is not saturated. If a parcel of air at the surface were to be lifted, its temperature will decrease according to the dry adiabatic lapse rate up to the LCL. In our example in Figure 8, the LCL is just above the 900 mb level. If the parcel continued to rise above the LCL, its temperature will decrease according to the saturated adiabatic lapse rate. The dashed line on the diagram in Figure 8 shows how the parcel temperature changes with height starting at the surface.

In Figure 8, there is a level between 700 mb and 800 mb where the parcel temperature and environmental temperature are equal (the solid and dashed lines intersect). This level (or altitude) is called the **Level of Free Convection** or LFC.

5. In Figure 8, compare the parcel temperature with the environmental temperature at several altitudes between the surface and the LFC. Is the parcel stable, neutral, or unstable?

6. At the LFC, is the parcel stable, neutral, or unstable?

7. Between 200 mb and 300 mb, the parcel temperature and environmental temperature are once again equal. That's where the solid and dashed lines intersect. Between that altitude and the LFC, is the parcel stable, neutral, or unstable? (Compare the parcel temperature and environmental temperature at several different altitudes.)

8. If parcels of air could be lifted from the surface to the LCL, a cloud would begin to form. If the parcels continued to be forced upward to just above the LFC, would you expect a stratus cloud or a cumulonimbus cloud to form?

III. SOME ADDITIONAL CLOUD TYPES

Clouds can be categorized into four major groups according to the altitude at which they form. In this section, we will focus on the high cloud group that includes three types of clouds: cirrus, cirrostratus, and cirrocumulus. In addition, we will also describe a special type of cloud called a contrail.

A. CIRRUS CLOUDS

Clouds that form in the upper part of the troposphere where temperatures are very cold are composed of tiny ice crystals. One type of high cloud is called *cirrus*. An example is seen in Figure 9.

FIGURE 9

Cirrus Cloud

Image © jennyt, 2014. Used under license from Shutterstock, Inc.

B. CIRROSTRATUS CLOUDS

Another type of high cloud is called *cirrostratus*. Cirrostratus clouds often cover most or all of the sky. When they are thin, a halo can sometimes be seen around the sun or moon. Figure 10 shows a halo around the sun. This occurs when light is refracted as it passes through the ice crystals of a cirrostratus cloud.

FIGURE 10

Cirrostratus Cloud with Halo Around the Sun

Thicker cirrostratus clouds often move overhead in advance of an approaching storm system.

C. CIRROCUMULUS CLOUDS

The third type of high cloud is called *cirrocumulus* (Figure 11).

FIGURE 11

Cirrocumulus Cloud

Cirrocumulus clouds often exhibit some vertical development.

9. Which type of cloud, cirrostratus or cirrocumulus, is associated with instability?

D. CONTRAIL

A type of cloud that forms behind jet aircraft flying at high altitudes is called a condensation trail or *contrail* (Figure 12). Contrails form from the mixing of water vapor from the exhaust of jet engines with cold air.

FIGURE 12

Contrail Behind a Jet Aircraft

Name: _____

Course Number _____ Section Number _____

ANSWER SHEET FOR LAB 6

1. In Figure 5, if the parcel were forced from 100 meters up to 200 meters, what would the parcel temperature be at 200 meters?

2. Is the parcel stable or unstable at 200 meters?

3. In Figure 6, as the parcel continues to rise from 100 meters up to 200 meters, what is the parcel temperature at 200 meters?

4. Is the parcel stable or unstable at 200 meters?

5. In Figure 8, compare the parcel temperature with the environmental temperature at several altitudes between the surface and the LFC. Is the parcel stable, neutral, or unstable?

6. In Figure 8, at the LFC, is the parcel stable, neutral, or unstable?

7. In Figure 8, between 200 and 300 mb, the parcel temperature and environmental temperature are once again equal. That's where the solid and dashed lines intersect. Between that altitude and the LFC, is the parcel stable, neutral, or unstable? (Compare the parcel temperature and environmental temperature at several different altitudes.)

8. If parcels of air could be lifted from the surface to the LCL, a cloud would begin to form. If the parcels continued to be forced upward to just above the LFC, would you expect a stratus cloud or a cumulonimbus cloud to form?

9. Which type of cloud, cirrostratus or cirrocumulus, is associated with instability?

LAB 7

AIR MASSES AND FRONTS

I. INTRODUCTION

A. OVERVIEW

One of the factors that can cause changes in the weather we observe is the movement of air masses. In this lab, we are going to describe some different types of air masses and look at a feature associated with air masses called a **front**. We will explain how fronts, such as cold fronts and warm fronts, are identified on surface weather maps. The **dryline** will also be discussed.

B. OBJECTIVES

Upon completion of this lab, you should be able to:
1. Identify four types of air masses.
2. Locate the position of a cold front using surface weather observations.
3. Explain what a dryline is and how to locate its position on a surface weather map.

II. AIR MASSES

An **air mass** is a very large body of air with similar horizontal temperature and moisture properties. The temperature and moisture properties of an air mass are acquired when a large body of air remains over a certain geographical region of the earth for a period of time. This geographical region is called a **source region**. The longer an air mass remains over its source region, the more the air mass takes on the characteristics of the underlying surface in terms of temperature and moisture.

Four types of air masses will be described in this section. These are: continental polar (cP), continental arctic (cA), maritime tropical (mT), and continental tropical (cT). The terms continental and maritime refer to the moisture content of the air mass. A continental air mass forms over land, so the air contains a relatively low amount of water vapor. These air masses are termed *dry* because of their low moisture content.

A maritime air mass forms over water, so it has a relatively high amount of water vapor. Maritime air masses are termed *moist*.

Polar, arctic, and tropical are terms associated with the temperature property of an air mass. A polar or arctic air mass is relatively cold compared to a tropical air mass that is warm. An arctic air mass is characterized by bitterly cold temperatures.

The source region for continental arctic air masses is the Arctic area of northern Canada. Continental polar air masses form over Alaska and parts of southern Canada. One of the major source regions for maritime tropical air masses is the Gulf of Mexico. Lastly, continental tropical air masses form over the desert regions of Old Mexico and southern New Mexico.

1. Use Figure 1 showing four air mass source regions to fill in Table 1. Terms to be used for the temperature characteristic are *warm*, *cold*, and *very cold*. Terms to be used for the moisture characteristic are *dry* and *moist*.

TABLE 1

Source Region	Name of Air Mass	Moisture Characteristic	Temperature Characteristic
1			
2			
3			
4			

FIGURE 1

Air Mass Source Regions

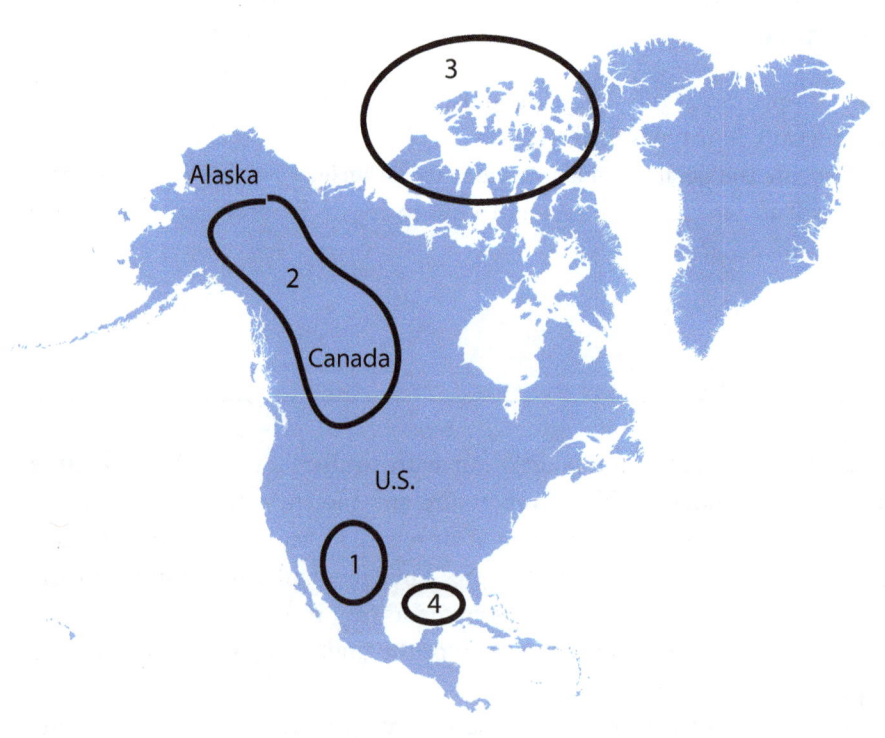

III. FRONTS

A. DESCRIPTION

Air masses don't remain over their source region for an indefinite period of time. As they move from their source region, one air mass can meet another. For example, a continental polar air mass moving south can meet a maritime tropical air mass moving north into the U.S. When air masses come together, they don't readily mix to create a new type of air mass. This is because of the different densities of the air masses. Cold air is more dense than warm air. The warm air is more buoyant and rises above the cold air. A boundary, or transition zone, exists between the air masses. This boundary is called a front. Depending on the way air masses move, a front can be labeled cold, warm, stationary, or occluded. In this lab, we will only consider cold and warm fronts.

Figure 2 is a conceptual model of a mid-latitude cyclone (surface low pressure center). Also shown is a cold front, identified with triangles, and a warm front, identified with half circles.

Map outline image © chrupka, 2014. Used under license from Shutterstock, Inc.

FIGURE 2

Conceptual model of a mid-latitude cyclone with fronts.

The counter-clockwise circulation of air around a low pressure center causes a continental polar or continental arctic air mass to advance southward on the west side of the low. The leading edge of the advancing cold air mass at the surface is marked by the cold front. The symbol for a cold front looks like this:

The triangles point in the direction the front is moving.

East of the low pressure center, southerly winds bring warm, moist air northward from the Gulf of Mexico. The boundary between this advancing warm air and the relatively colder air farther north is the warm front. The symbol for a warm front looks like this:

Similar to the cold front, the half-circles indicate the direction the warm front is moving.

B. LOCATING FRONTS

A frontal passage at a particular location can be associated with a rapid change in temperature, wind, and humidity. In addition, fronts are features in the atmosphere associated with rising air. The warmer, less dense air is lifted above the colder, denser air. If sufficient moisture and lift are present, clouds and precipitation form.

Since a variety of weather changes can be associated with fronts, it's important to identify their location. To locate the position of a cold or warm front on a surface weather map, one should keep in mind the characteristic properties of the different air masses and the conceptual model shown in Figure 2.

Figure 3 is a surface weather map for a particular time. To locate a front, we need observations of temperature, dew point, and wind. Recall that dew point is an indicator of the amount of moisture (water vapor) in the air. The higher the dew point, the greater the amount of water vapor. Sea level pressure and isobars drawn to sea level pressure are important, but we will not consider pressure for this lab.

To facilitate plotting weather observations at different locations, a format called a **station model** is used. A simple station model used on a surface weather map has the format shown below:

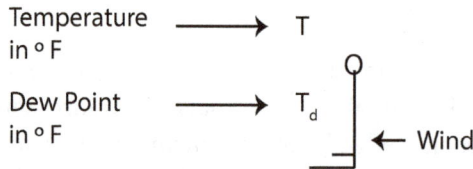

Figure 4 is an example of a station model based on observations of temperature, dew point, and wind taken at a particular place and time.

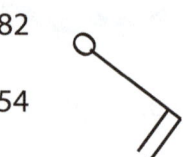

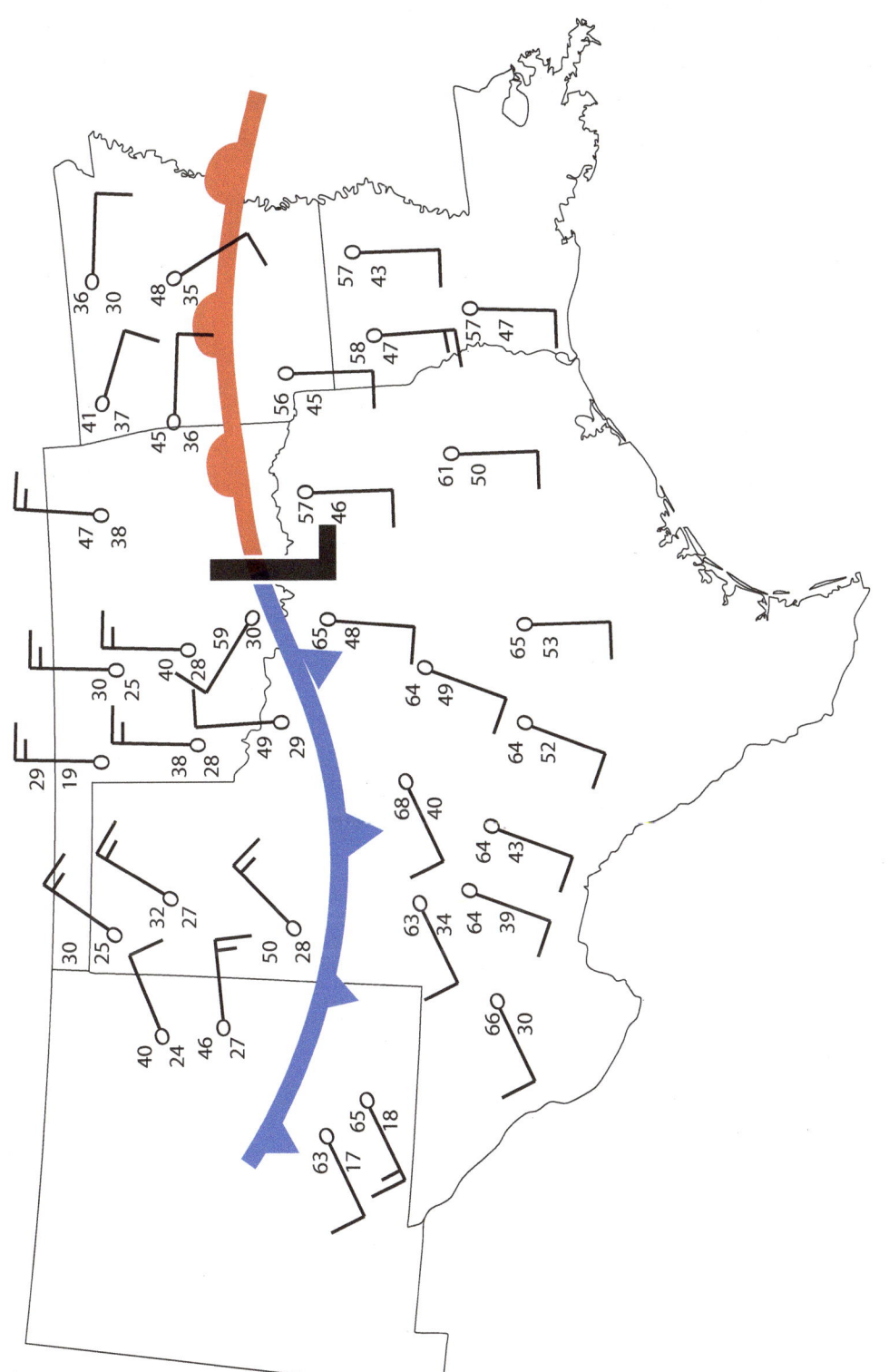

FIGURE 3

Surface Weather Map

Map outline image © chrupka, 2014. Used under license from Shutterstock, Inc.

In Figure 4, the temperature is 82°F. The dew point is 54°F. Interpreting the wind symbols plotted on weather maps was explained in Lab 1. As a refresher, the circle marks the location of the particular station where the observations were made. The line extending from the station circle denotes the wind direction. Wind direction is plotted according to the direction *from* which is the air is coming. The smaller lines (called "feathers") denote the wind speed. A full line represents 10 knots and a half line represents 5 knots. In Figure 4, the wind direction is from the southeast and the wind speed is 20 knots.

A wind speed of around 5 knots is shown by the following:

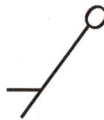

2. Decode the following station models on the answer sheet:

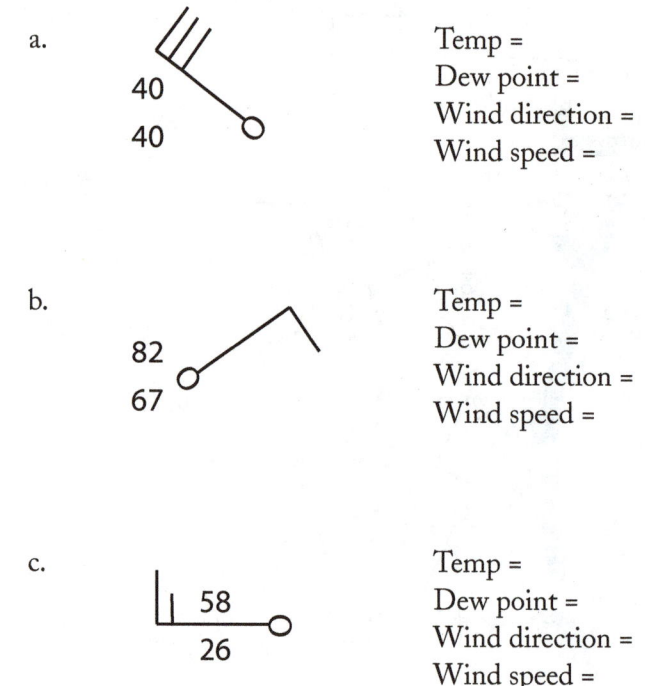

a.

Temp =
Dew point =
Wind direction =
Wind speed =

b.

Temp =
Dew point =
Wind direction =
Wind speed =

c.

Temp =
Dew point =
Wind direction =
Wind speed =

3. Using the following observations, draw a station model for each city on the answer sheet.

City A: Temp = 68°F
   Dew point = 47°F
   Wind is south at 15 knots

O

City B: Temp = 91°F
   Dew point = 33°F
   Wind is southwest at 20 knots

O

City C: Temp = 32°F
    Dew point = 31°F   O
    Wind is north at 35 knots

   We can now discuss why the fronts shown in Figure 3 were drawn the way you see them. Use Figure 3 to answer questions 4 through 9. These questions should guide you to an understanding of how to locate fronts on a surface weather map.

4. Is the colder air mass north or south of the cold front?

5. Compare the dew point temperatures north of the cold front to those south of the cold front in Figure 3. Is the air mass north of the cold front relatively dry or moist compared to south of the cold front?

6. The air mass north of the cold front is
    A. continental tropical
    B. continental polar
    C. maritime tropical

7. The air mass south of the cold front is
    A. continental tropical
    B. continental polar
    C. maritime tropical

8. In general, the wind direction north of the cold front is from the
    A. north
    B. south

9. In general, the wind direction south of the cold front is from the
    A. north
    B. south

   The cold front is drawn to show the separation, or boundary, between a colder, drier air mass and a warmer, moister air mass. Temperatures, dew points, and wind directions are used in locating the front.
   Note the wind directions north of the warm front. They are predominantly from the east or southeast. This indicates the cold air mass north of the warm front is retreating to the north. This will allow the warm air mass south of the warm front to move toward the north. The warm front is the leading edge of the northward advancing warmer air mass.

10. Figure 5 on page 103 is a surface weather map showing plotted observations of temperature, dew point, wind speed, and wind direction for a particular day in December at 6:00 pm CST. A portion of a cold front has already been

drawn on this map. Using the concepts previously discussed, complete the drawing of the cold front on Figure 5.

11.   Figure 6 on page 105 is a surface weather map similar to Figure 5 except the observations plotted were taken 12 hours later, at 6:00 am CST. During this 12-hour period, the cold front moved east and south. Complete the drawing of the cold front on Figure 6.

IV. THE DRYLINE

A. DESCRIPTION

A type of air mass boundary commonly found in the southern plains states of Kansas, Oklahoma, and Texas during the spring and early summer is called a dryline. The dryline separates a continental tropical air mass from a maritime tropical air mass.

There is generally very little temperature contrast between a continental tropical and a maritime tropical air mass. The difference between the two air masses is in the moisture content. Continental tropical air masses contain a relatively low amount of water vapor. On the other hand, maritime tropical air masses contain higher amounts of water vapor.

B. LOCATING A DRYLINE

To locate the position of a dryline using surface weather observations plotted on a map, one primarily focuses on dew point temperatures and wind directions. Look at Figure 7 and note the position of the dryline.

The symbol for a dryline is a line with scallops as in the following example:

You should observe in Figure 7 that air temperatures west of the dryline are not that much different from air temperatures east of it. Answering questions 12 through 15 should lead you to an understanding of why the dryline was drawn the way it is in Figure 7.

12.   Are dew points east of the dryline higher or lower than west of the dryline?

13.   Does the air mass east of the dryline contain more or less water vapor (moisture) than the air mass west of the dryline?

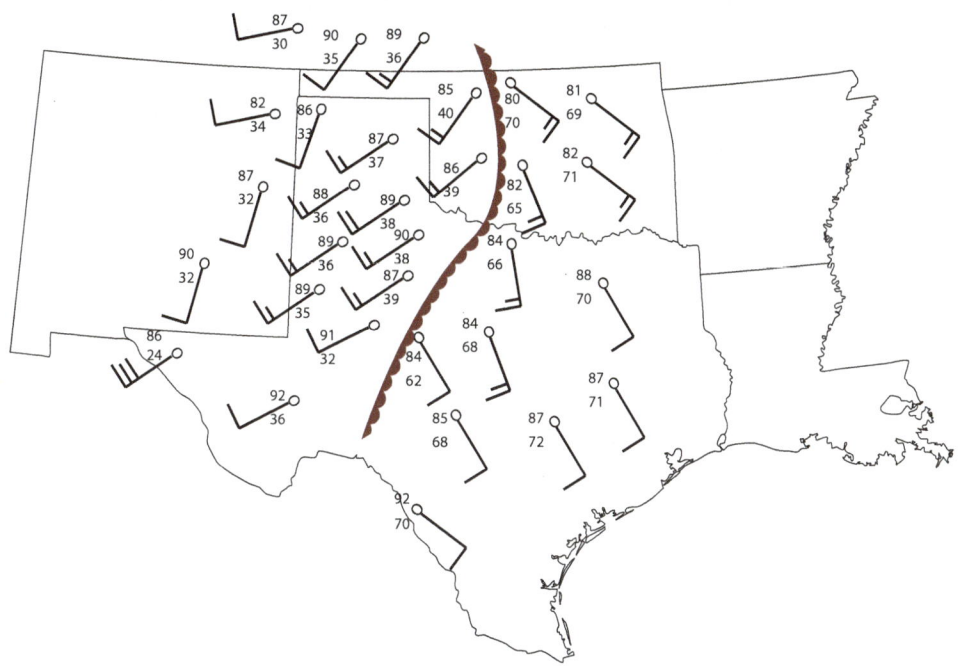

14. The type of air mass west of the dryline is
    A. maritime tropical
    B. continental tropical
    C. continental polar

15. The type of air mass east of the dryline is
    A. maritime tropical
    B. continental tropical
    C. continental polar

Wind directions are also used in locating the dryline. Winds east of the dryline are typically from the south or southeast, while west of the dryline winds are from the southwest or west.

16. The observations plotted on the map in Figure 8 on page 107 reveal a cold front had moved southward into Arkansas, Oklahoma, and the Texas Panhandle.

    a. On Figure 8, draw the cold front. Then draw the dryline. You should start drawing the dryline at some point on the cold front.
    b. Correctly label the three air masses on the map. The three air masses are continental polar (cP), continental tropical (cT), and maritime tropical (mT).

There are instances, such as shown in Figure 8, where a cold front and dryline intersect. This is called the *triple point* and is sometimes a favored area for thunderstorm development.

C. IMPORTANCE OF THE DRYLINE

The dryline is a very important meteorological feature because it is often the focus for thunderstorm development. This will be discussed in Lab 8.

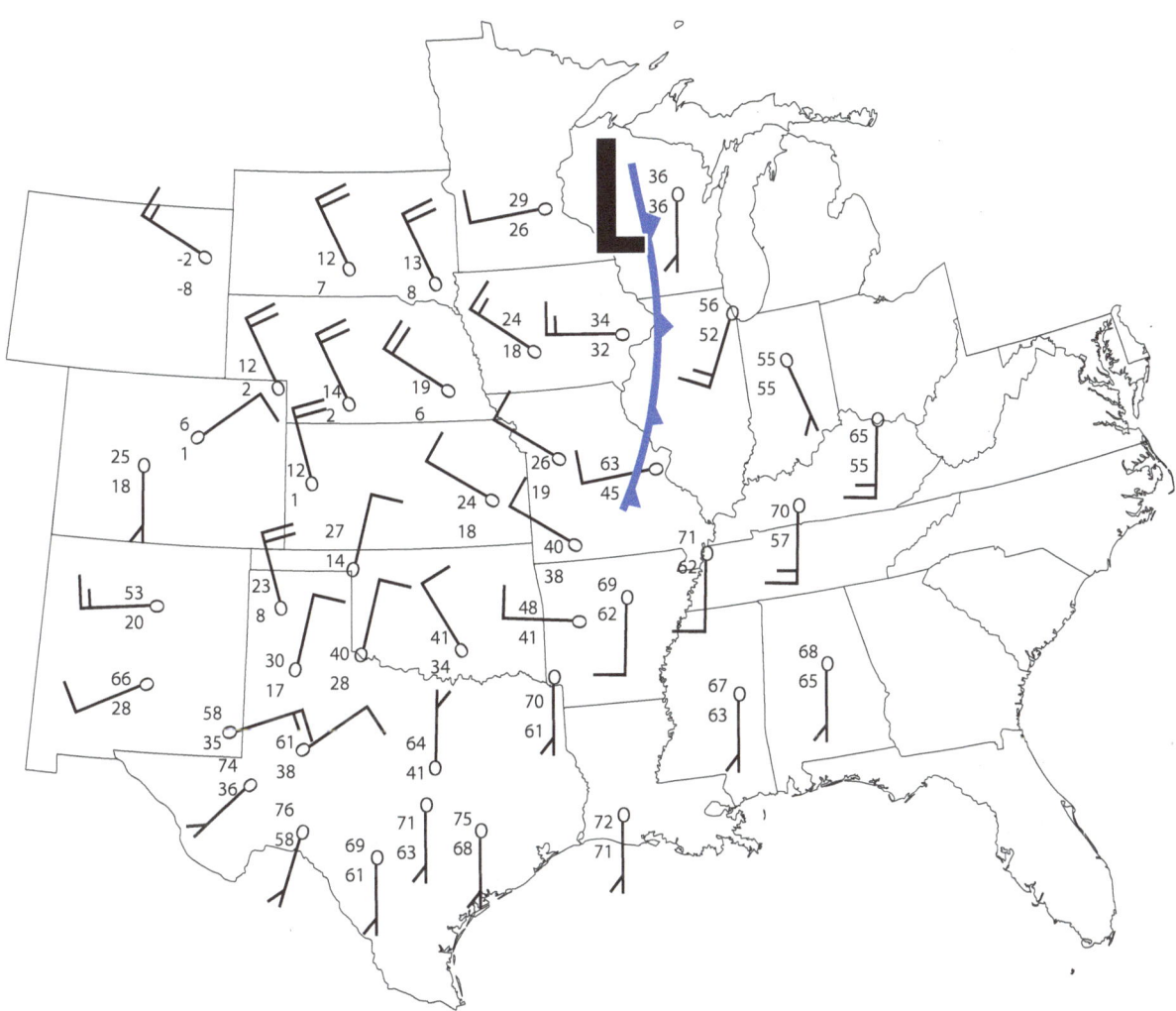

FIGURE   5

*Surface Weather Map
for 4 December at
6 pm CST*

Map outline image © chrupka, 2014. Used under license from Shutterstock, Inc.

FIGURE 6

*Surface weather map
for 5 December at
6am CST*

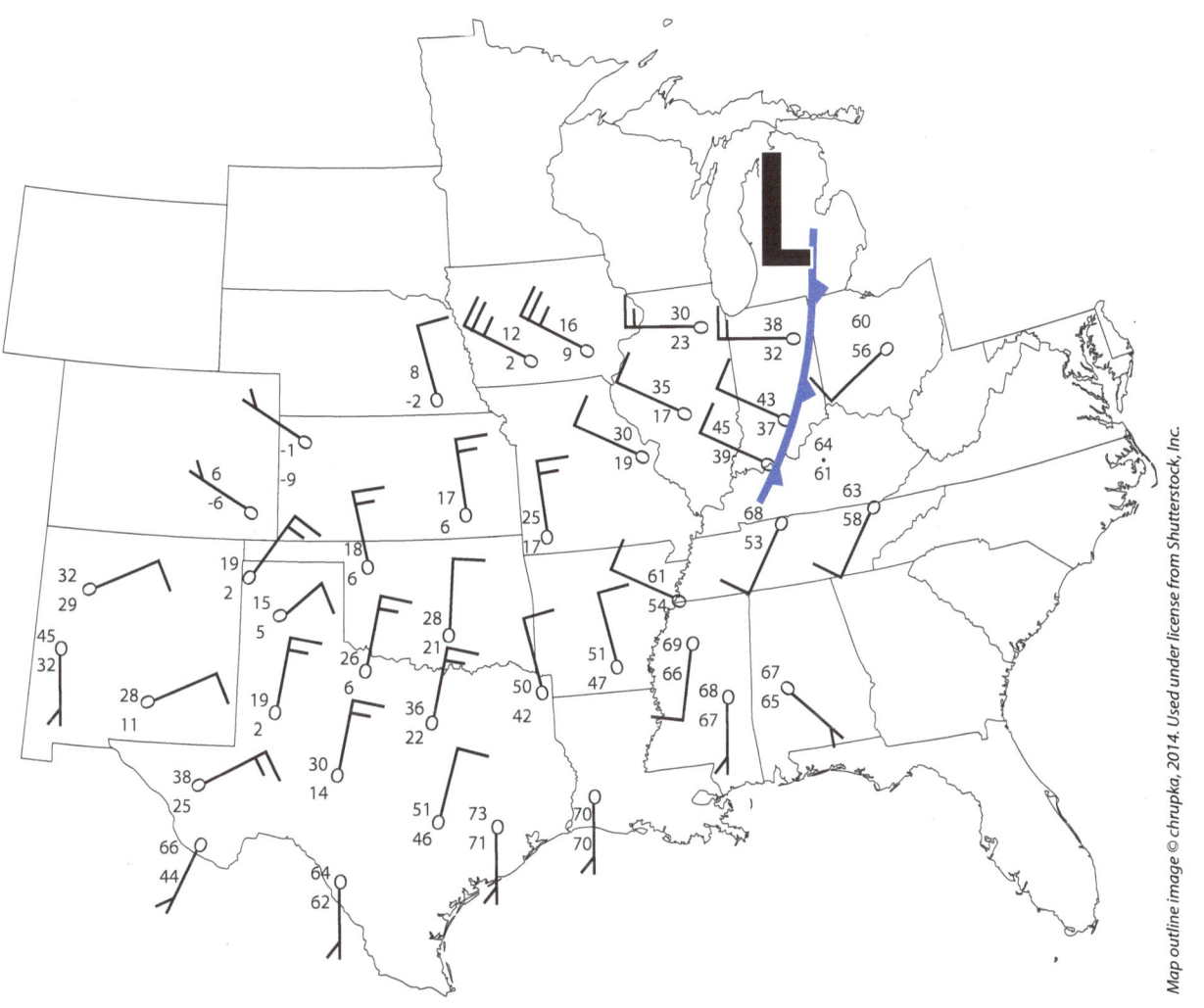

FIGURE 8

*Surface
Weather Map*

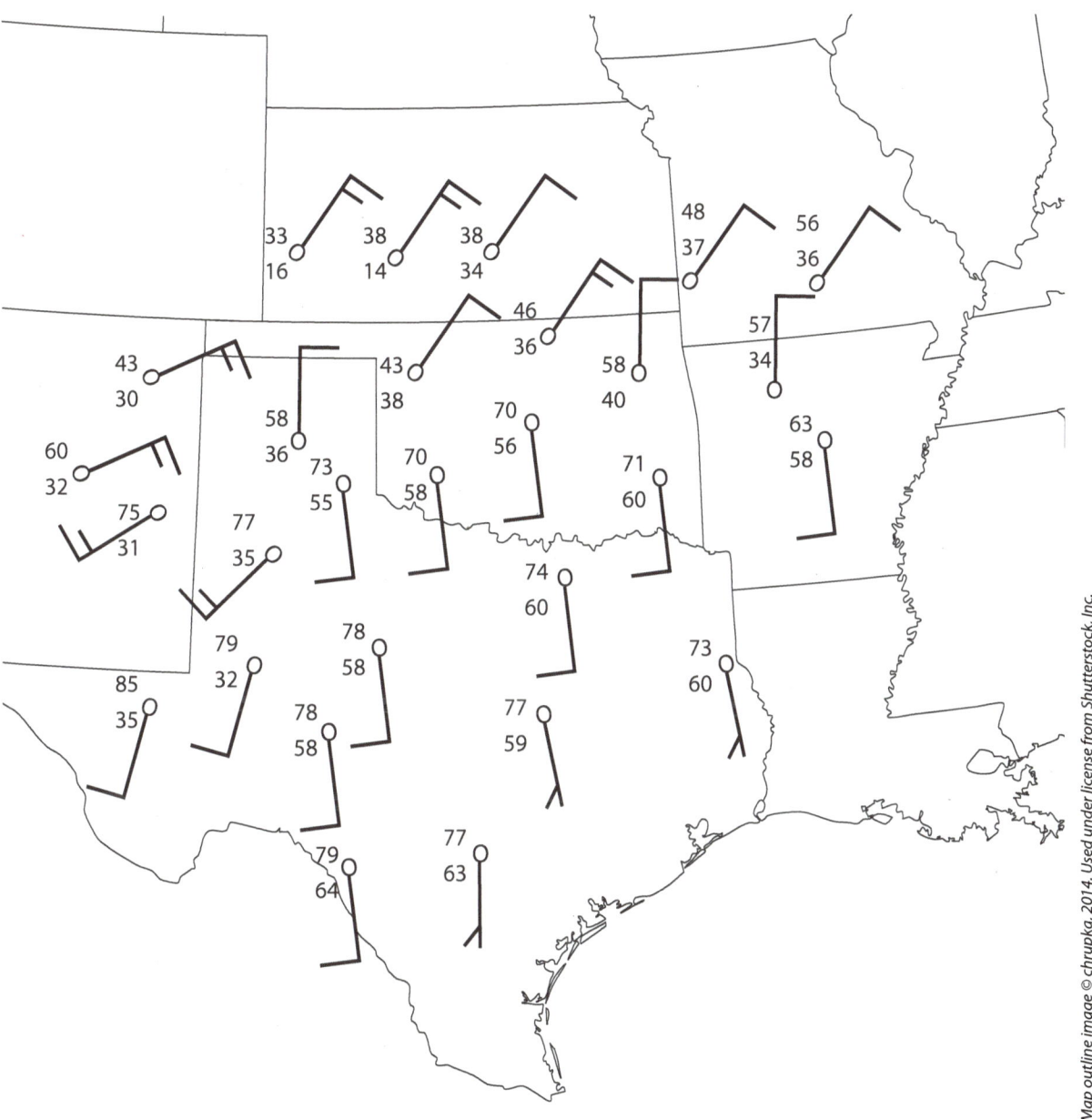

Name: _____

Course Number _____ Section Number _____

ANSWER SHEET FOR LAB 7

1. Use Figure 1 showing four air mass source regions to fill in Table 1 below. Terms to be used for the temperature characteristic are *warm*, *cold*, and *very cold*. Terms to be used for the moisture characteristic are *dry* and *moist*.

Source Region	Name of Air Mass	Moisture Characteristic	Temperature Characteristic
1			
2			
3			
4			

2. Decode the following station models:
   a. Temp =
      Dew point =
      Wind direction =
      Wind speed =

   b. Temp =
      Dew point =
      Wind direction =
      Wind speed =

   c. Temp =
      Dew point =
      Wind direction =
      Wind speed =

3. Using the following observations, draw a station model for each city.

   City A: Temp = 68°F
   Dew point = 47°F   O
   Wind is south at 15 knots

City B: Temp = 91°F
Dew point = 33°F   O
Wind is southwest at 20 knots

City C: Temp = 32°F
Dew point = 31°F   O
Wind is north at 35 knots

Use Figure 3 to answer questions 4 through 9.

4. Is the colder air mass north or south of the cold front?

5. Compare the dew point temperatures north of the cold front to those south of the cold front. Is the air mass north of the cold front relatively dry or moist compared to south of the cold front?

6. The air mass north of the cold front is
    A. continental tropical
    B. continental polar
    C. maritime tropical

7. The air mass south of the cold front is
    A. continental tropical
    B. continental polar
    C. maritime tropical

8. In general, the wind direction north of the cold front is from the
    A. north
    B. south

9. In general, the wind direction south of the cold front is from the
    A. north
    B. south

10. Figure 5 is a surface weather map showing plotted observations of temperature, dew point, wind speed, and wind direction for 4 December at 6:00 pm CST. A portion of a cold front has already been drawn on this map. Using the concepts previously discussed, complete the drawing of the front on Figure 5.

11. Figure 6 is a surface weather map similar to Figure 5 except the observations plotted were taken 12 hours later, on 5 December at 6:00 am CST. During this 12-hour period, the cold front moved east and south. Complete the drawing of the front on Figure 6.

Use Figure 7 to answer questions 12 through 15.

12. Are dew points east of the dryline higher or lower than west of the dryline?

13. Does the air mass east of the dryline contain more or less water vapor (moisture) than the air mass west of the dryline?

14. The type of air mass west of the dryline is
    A. maritime tropical
    B. continental tropical
    C. continental polar

15. The type of air mass east of the dryline is
    A. maritime tropical
    B. continental tropical
    C. continental polar

16. The observations plotted on the map in Figure 8 reveal a cold front had moved southward into Arkansas, Oklahoma, and the Texas Panhandle.

    a. On Figure 8, draw the cold front. Then, starting at some point on the cold front, draw the dryline.

    b. Correctly label the three air masses on the map. The three air masses are continental polar (cP), continental tropical (cT), and maritime tropical (mT).

LAB 8

THUNDERSTORMS AND DOPPLER RADAR

I. INTRODUCTION

A. OVERVIEW

In this lab, we will discuss the primary weather ingredients needed for thunderstorm formation. Recorded weather observations taken on a particular day will be used to identify those ingredients. We will focus on one type of thunderstorm called a supercell.

One of the tools used to observe thunderstorms is Doppler radar. We will present a very limited description of how radar works and look at one of the images produced by the radar called the reflectivity image.

B. OBJECTIVES

Upon completion of this lab, you should be able to:
1. Identify atmospheric conditions favorable for the formation of supercell thunderstorms.
2. Name the criteria for a thunderstorm to be classified as severe.
3. Interpret a reflectivity radar image.

II. ATMOSPHERIC CONDITIONS FAVORABLE FOR THUNDERSTORMS

Thunderstorms form where there is sufficient low-level moisture (water vapor), a mechanism to initially lift the air so the air can cool to the saturation point (the LCL), and instability (unstable parcels of air). These ingredients have been discussed in previous labs, but a brief review will be provided here.

The initial formation of a cumulonimbus cloud occurs when water vapor in the air condenses into small liquid water droplets. For the condensation process to begin, the air must become saturated. As discussed in Lab 5, the

primary way for the air to become saturated is cooling the air. The cooling comes about when air rises.

Some of the features in the atmosphere associated with rising air include:
  A. Convergence of winds around a low pressure system at the surface (Lab 4)
  B. Mountains forcing air to rise (Lab 5)
  C. Air mass boundaries; e.g., cold fronts, warm fronts, and drylines (Lab 7)

Once a cloud begins to form, the future development of the cloud largely depends on whether or not the air is stable or unstable. This was discussed in Lab 6. For thunderstorms to develop, the rising parcels of air must be unstable. This means the air parcels have positive buoyancy that allows the parcels to accelerate upward through the troposphere.

III. SEVERE THUNDERSTORMS

A severe thunderstorm is one that produces one or more of the following:
  A. Hail one inch in diameter or larger
  B. Winds of 50 knots (58 mph) or higher
  C. Tornado

The wind criterion refers to what is called straight-line winds. These are the outflow winds that spread outward, away from the thunderstorm.

If either A or B above is observed or indicated by radar, the National Weather Service (NWS) will issue a severe thunderstorm warning to alert the public to seek shelter. If a tornado is observed or indicated by radar, the NWS will issue a tornado warning rather than a severe thunderstorm warning. Notice that neither heavy rain nor lightning is a criterion for classifying a thunderstorm as severe.

One type of thunderstorm that will be emphasized in this lab is called a **supercell**. A supercell storm can produce very large hail, strong and damaging outflow winds, and tornadoes. The destructive nature of supercell thunderstorms is due, in part, to rotation of air in the updraft region of the storm. The rotating updraft is the distinguishing characteristic of a supercell.

In addition to the three ingredients needed for thunderstorms to form, a supercell requires the presence of **vertical wind shear** in the troposphere. Vertical wind shear refers to the change in the horizontal wind with height. It could be the wind speed increasing with height or the direction of the wind changing with height or both.

A detailed explanation of the role of vertical wind shear in the formation of supercell thunderstorms is beyond the purpose of this lab. Simply stated, vertical wind shear helps to create the rotating updraft.

1. a. Assume you are working in a NWS office in some city. A thunderstorm has developed close to the airport in your city. The anemometer measures a wind

gust of 53 mph and you believe this wind gust is outflow winds from the thunderstorm. Based on this, you would

A. issue a tornado warning

B. issue a severe thunderstorm warning

C. issue no warning

b. Five minutes later, a reliable weather observer calls the NWS office and informs you that a thunderstorm 15 miles southwest of the city is producing hail up to 1.25 inches in diameter. Based on this report, you would

A. issue a tornado warning

B. issue a severe thunderstorm warning

C. issue no warning

IV. FORECASTING SEVERE THUNDERSTORMS

Weather forecasters determine the potential for severe thunderstorm development by analyzing the four ingredients previously discussed (moisture, lift, instability, and vertical wind shear). Forecasters first analyze the current state of the atmosphere in terms of those four ingredients. Then they must determine how the current state of the atmosphere will change over time. To do this, they employ computer models of the atmosphere known as numerical weather prediction.

In this section, we will use actual weather observations taken on a day in June to describe the process of analyzing the four ingredients for severe thunderstorms.

A. WATER VAPOR (MOISTURE)

A method to evaluate the availability of low-level moisture is to look at dew points on a surface weather map. Generally speaking, severe storm forecasters look for surface dew points of 50°F to 55°F or higher. (Thunderstorms can occur when surface dew points are quite low. In fact, thunderstorms can occur when it's snowing. This is called thundersnow.)

Figure 1 on page 123 is a surface weather map depicting weather observations taken at 5:00 pm CDT on June 9. Use this figure to answer questions 2 through 8.

2. What is the dew point at Lubbock, Texas (LBB)?

3. What is the dew point at Clovis, New Mexico (CVS), northwest of Lubbock?

4. Which location, Lubbock or Clovis, has the greater amount of water vapor in the air?

Without making any notations on the map, just observe where dew points are high and where they are low.

B. LIFTING MECHANISM

As stated earlier, several mechanisms can initially force air to rise, leading to saturation and cloud formation. For this lab, we will only consider the dryline that was discussed in Lab 7.

5. On Figure 1, draw in the location of the dryline.

6. What is the predominant wind direction at the observation sites east of the dryline?

7. What is the predominant wind direction at the observation sites west of the dryline?

These opposing wind directions indicate air is coming together at the dryline. This is called convergence and is associated with rising air.

8. Based on where the highest amount of moisture is in relation to the dryline, would you expect the greater likelihood for thunderstorm formation to be
    A. along and just west of the dryline
    B. along and just east of the dryline

C. INSTABILITY

Recall from Lab 6 that the stability of air parcels, in effect, whether they are stable or unstable, is determined by comparing the parcel temperature with the environmental temperature. In Lab 6, we illustrated that by looking at a diagram showing air temperatures measured by a radiosonde at different altitudes along with the temperature in a rising parcel of air. Figure 2 shows that diagram from Lab 6.

FIGURE 2

Environmental (solid line) and Parcel (dashed line) Temperatures

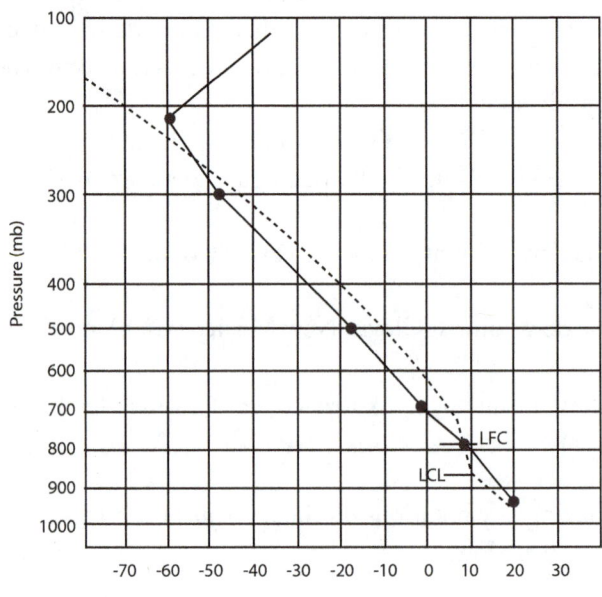

LAB 8 · *Thunderstorms and Doppler Radar*

Notice in Figure 2 that at each altitude from the LFC (Level of Free Convection) upward to just below the 200 mb level, the parcel temperature is warmer than the environmental temperature. Therefore, the parcel is unstable and will rise due to buoyancy.

If numerous air parcels can be forced upward from the surface to just above the LFC, the parcels will continue accelerating upward to form the cumulonimbus cloud. The more unstable the parcels are, the more positive buoyancy the parcels have. This means the updraft will be stronger. This is important because supercell thunderstorms generally form in an atmosphere where the parcels are very unstable. Large hail that often forms in a supercell is associated with very strong updrafts. In some supercell storms, updraft speeds can reach 100 mph or even higher.

On June 9, a radiosonde observation was made near Lubbock around 1:00 pm. Normally radiosonde observations are not made at Lubbock but it was done on this day to support research and forecast operations. Figure 3 shows a plot of temperature readings taken by this radiosonde at various altitudes (solid line). Also shown in Figure 3 are temperatures in a rising parcel of air that were computed based on the dry and saturated adiabatic lapse rates (dashed line).

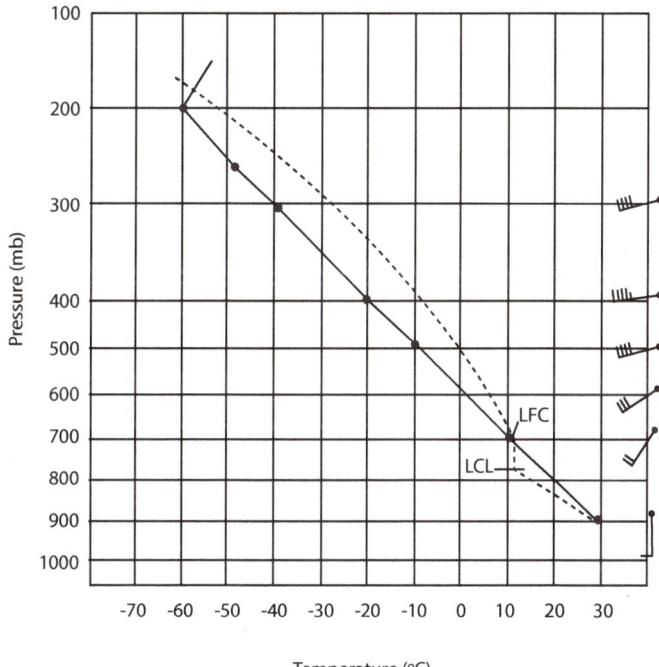

FIGURE 3

Radiosonde sounding near Lubbock, TX at 1:00 pm CDT on June 9. Solid line represents the environmental temperature. Dashed line represents the parcel temperature.

9. Using Figure 3, would a rising parcel of air be stable or unstable between the LFC and 200 mb?

10. Given this sounding in Figure 3, clouds would begin to form if parcels of air could be lifted from the surface to the LCL. The LCL on June 9 was between 750 mb and 800 mb. What weather feature at the surface on this day could force air parcels to rise to the LCL?

11. If parcels of air continued to be lifted to just above the LFC, would you expect a stratus cloud or a cumulonimbus cloud to develop?

D. VERTICAL WIND SHEAR

On the right side of Figure 3, the observed winds at several altitudes based on the radiosonde observation are shown. The symbols showing wind direction and wind speed are read just like the winds plotted on weather maps (refer to Labs 1 and 2). For example, the wind at 600 mb is from the southwest at 30 knots.

12. Do wind speeds increase, decrease, or remain the same starting at the surface and going up to around 400 mb?

13. Do wind directions stay the same or do they change between the surface and 400 mb?

E. INTERIM SUMMARY

The following questions are intended to help summarize what has been presented so far in this lab.

14. In addition to instability (unstable parcels of air), what are the other two ingredients needed for general thunderstorm development (not just supercells)?

15. Given your answer to question 14, briefly describe the existence of those two ingredients from the weather observations on June 9. In other words, what evidence is there in the data provided that shows the presence of those two ingredients?

16. If you were forecasting for the Lubbock area on June 9, would you
    A. forecast some clouds but no thunderstorms?
    B. forecast possible thunderstorms but no supercells?
    C. forecast possible supercell thunderstorms?

Give a reason for your answer.

V. DOPPLER RADAR

Once thunderstorms begin to develop over a particular area, meteorologists use Doppler radar to monitor the location, movement, and intensity of the thunderstorms. Radar uses electromagnetic radiation in the form of microwaves to detect precipitation such as rain and hail within thunderstorms.

One component of a radar system is the antenna. The antenna is housed in a protective cover called a radome (Figure 4).

Image © pedrosala, 2014. Used under license from Shutterstock, Inc.

FIGURE 4

Picture of a Radome

The antenna constantly rotates 360 degrees. While it is rotating, it sends out electromagnetic energy in very short bursts called pulses. If this energy encounters precipitation-sized particles, some of the energy is scattered or reflected back to the antenna. This is called an echo. The radar system processes this returned signal and ultimately produces an image called a reflectivity image. An example is shown in Figure 5. The colors on this image will be explained using Figure 6. A radar reflectivity image shows the location and intensity of precipitation.

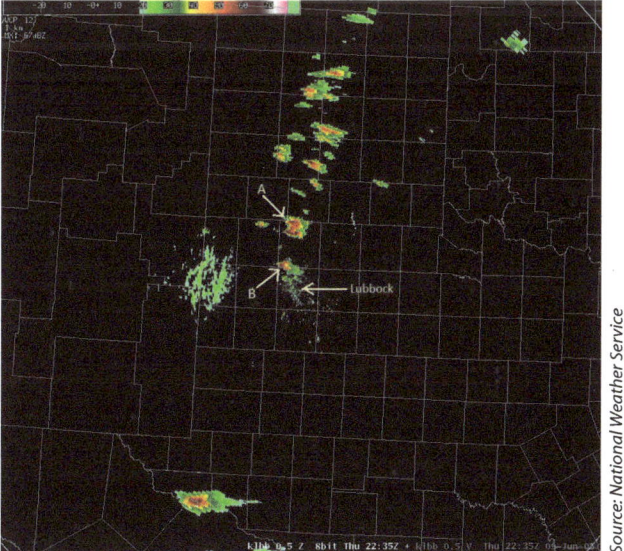

Source: National Weather Service

FIGURE 5

Reflectivity Image for 5:35 pm CDT on June 9

The reflectivity image in Figure 5 was made at 5:35 pm CDT on June 9.

Note the thunderstorms just north of Lubbock where arrows A and B are pointing.

Figure 6 is a reflectivity image for the same time but we have zoomed in on storms A and B north of Lubbock for a more detailed view.

FIGURE 6

Reflectivity Image for 5:35 pm CDT on June 9

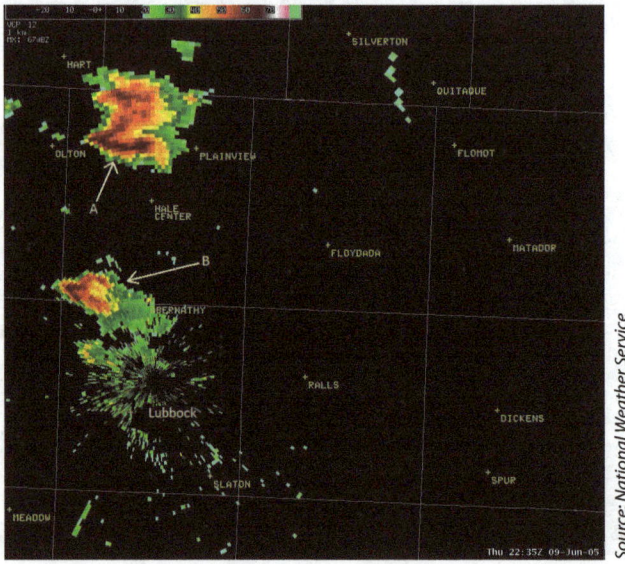

Source: National Weather Service

The green colors represent light to moderate precipitation. This means the returned signal (echo) was not very strong. The red and darker colors represent very intense rainfall or hail. In areas where there are many large raindrops and hail, more energy is returned to the antenna.

At 5:35 pm, storm A produced hail 1.00 inch in diameter. At the same time, storm B produced hail 1.75 inches in diameter (golf ball size).

17. Based on these reports, which storm(s) is/are classified as severe?

Figure 7 is the radar reflectivity image for 7:53 pm on June 9.

FIGURE 7

Reflectivity Image for 7:53 pm CDT on June 9

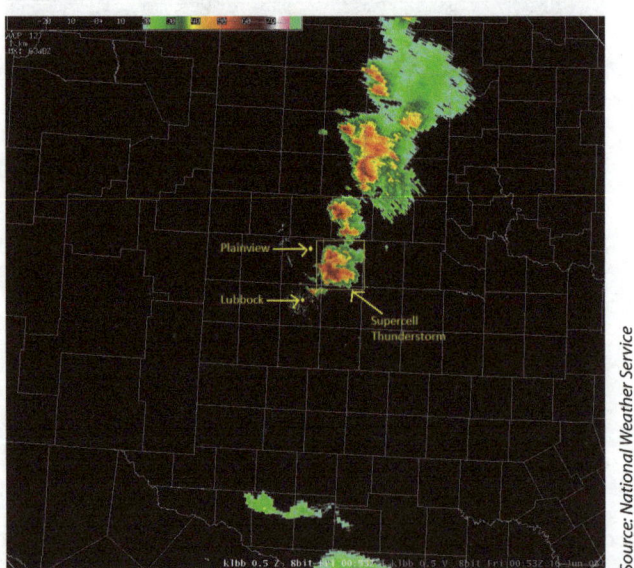

Source: National Weather Service

The thunderstorm northeast of Lubbock has indications that it is a supercell. To help identify the primary features associated with supercells as viewed on the reflectivity radar image, refer to the simplified drawing of a supercell below:

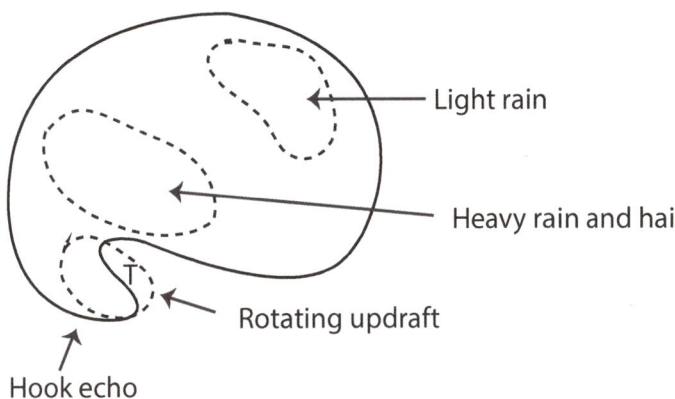

The key feature in identifying a supercell on the reflectivity image is the hook echo. This feature is associated with the rotating updraft. It should be emphasized that a hook echo does not necessarily mean a tornado is present. If a tornado does form, it typically will be located approximately where the letter "T" is in the diagram above.

Figure 8 is a zoomed in reflectivity image for 7:53 pm CDT on June 9 of the storm northeast of Lubbock.

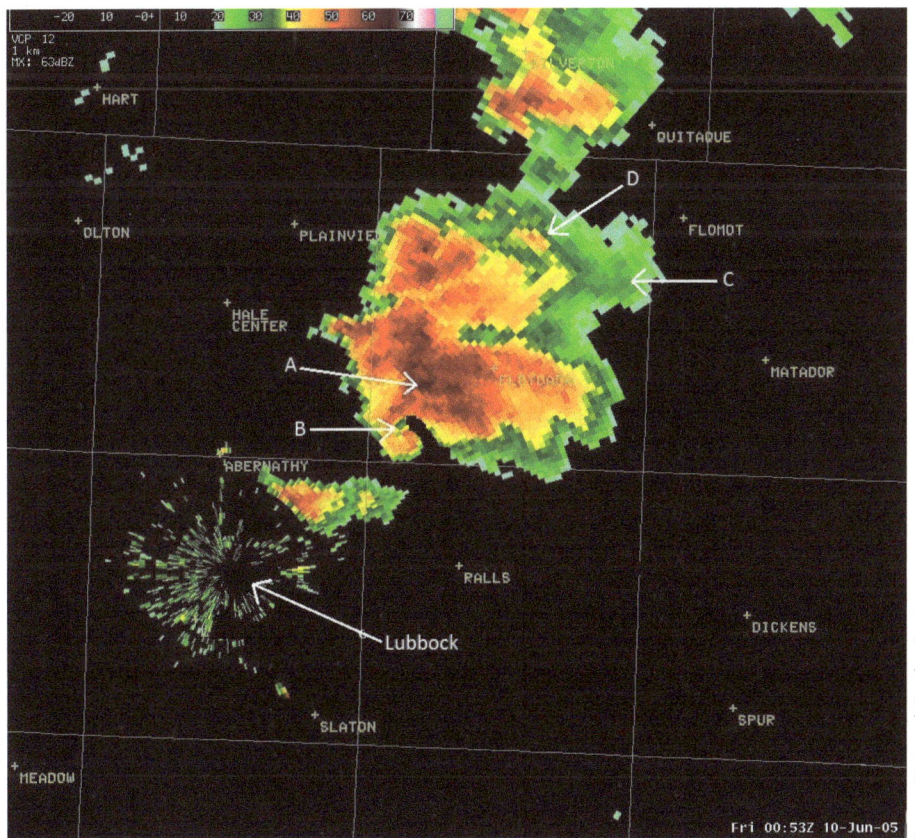

Source: National Weather Service

FIGURE 8

Reflectivity Image for 7:53 pm CDT on June 9

18. Which arrow in Figure 8 points to the location of the heaviest rain and hail?

19. Which arrow in Figure 8 points to the location of a possible tornado?

     A few tornadoes were reported with this storm along with golf ball sized hail (1.75 inches in diameter).

     In this lab, we only considered the reflectivity image from a Doppler radar. This is the type of image commonly seen on TV or the Internet.

     Doppler radar can also be used to observe air motion within thunderstorms. The image produced by the radar to show air motion is called the velocity image. An example is shown in Appendix B. The red colors denote air motion away from the radar antenna located at Lubbock. Blue and green colors denote air motion toward the radar antenna. This image coincides with the reflectivity image for 7:53 pm CDT on June 9 (Figure 8).

FIGURE 1

*Surface Weather Map
for 5:00 pm CDT
June 9th*

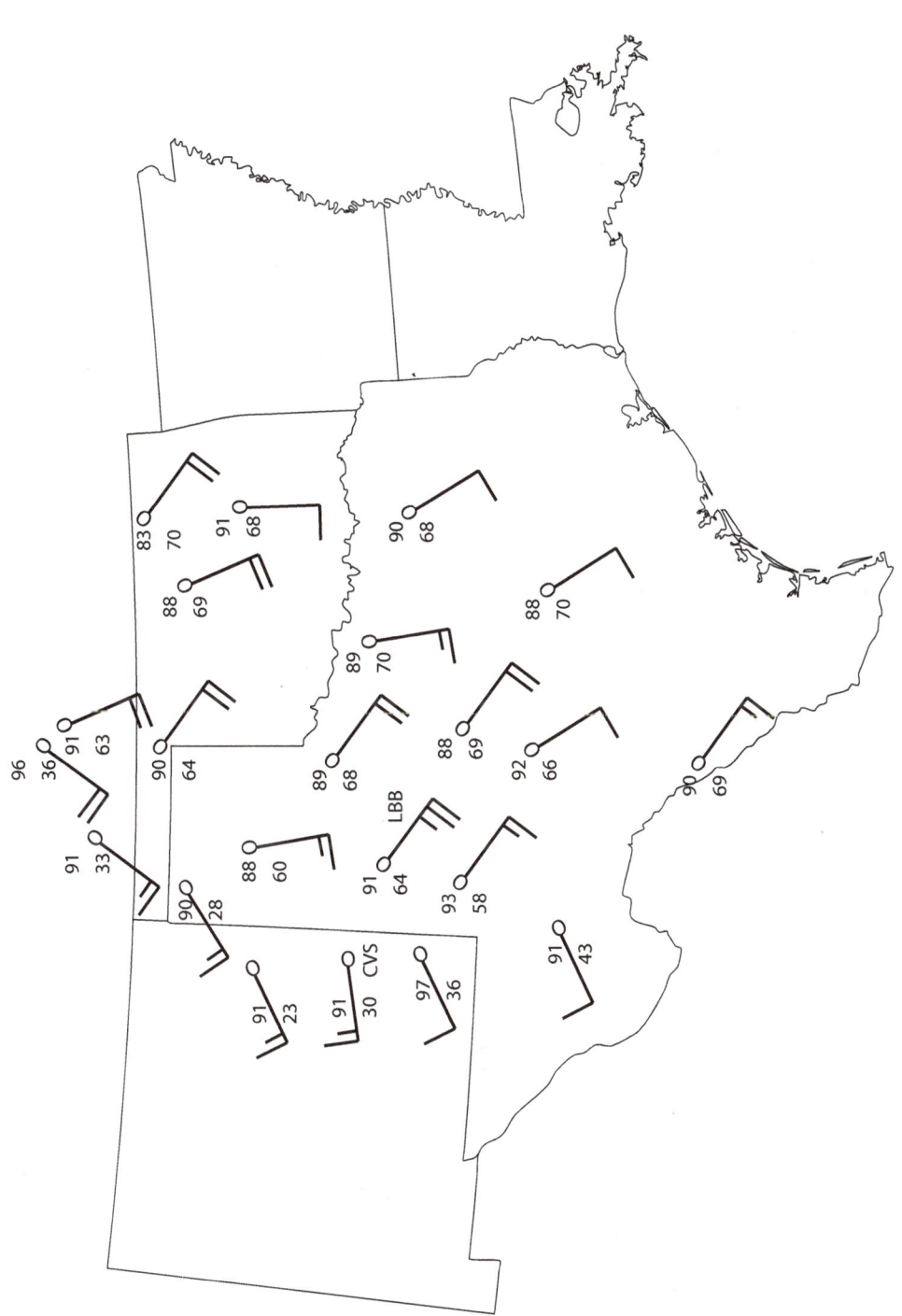

Map outline image © chrupka, 2014. Used under license from Shutterstock, Inc.

ANSWER SHEET FOR LAB 8

1. a. Assume you are working in a NWS office in some city. A thunderstorm has developed close to the airport in your city. The anemometer measures a wind gust of 53 mph and you believe this wind gust is outflow winds from the thunderstorm. Based on this, you would
    A. issue a tornado warning
    B. issue a severe thunderstorm warning
    C. issue no warning

   b. Five minutes later, a reliable weather observer calls the NWS office and informs you that a thunderstorm 15 miles southwest of the city is producing hail up to 1.25 inches in diameter. Based on this report, you would
    A. issue a tornado warning
    B. issue a severe thunderstorm warning
    C. issue no warning

Use Figure 1 on page 123 to answer questions 2 through 8.

2. What is the dew point at Lubbock, Texas (LBB)?

3. What is the dew point at Clovis, New Mexico (CVS), northwest of Lubbock?

4. Which location, Lubbock or Clovis, has the greater amount of water vapor in the air?

5. On Figure 1, draw in the location of the dryline.

6. What is the predominant wind direction at the observation sites east of the dryline?

7. What is the predominant wind direction at the observation sites west of the dryline?

8. Based on where the highest amount of moisture is in relation to the dryline, would you expect the greater likelihood for thunderstorm formation to be
   A. along and just west of the dryline
   B. along and just east of the dryline

Use Figure 3 to answer questions 9 through 13.

9. Using Figure 3, would a rising parcel of air be stable or unstable between the LFC and 200 mb?

10. Given this sounding in Figure 3, clouds would begin to form if parcels of air could be lifted from the surface to the LCL. The LCL on June 9 was between 750 mb and 800 mb. What weather feature at the surface on this day could force air parcels to rise to the LCL?

11. If parcels of air continued to be lifted to just above the LFC, would you expect a stratus cloud or a cumulonimbus cloud to develop?

12. Do wind speeds increase, decrease, or remain the same starting at the surface and going upward to around 400 mb?

13. Do wind directions stay the same or do they change between the surface and 400 mb?

14. In addition to instability (unstable parcels of air), what are the other two ingredients needed for general thunderstorm development (not just supercells)?

    A.

    B.

15. Given your answer to question 14, briefly describe in the space below the existence of those two ingredients from the weather observations on June 9. In other words, what evidence is there in the data provided on Figure 1 that shows the presence of those two ingredients?

    A.

    B.

16. If you were forecasting for the Lubbock area on June 9, would you
    A. forecast some clouds but no thunderstorms?
    B. forecast possible thunderstorms but no supercells?
    C. forecast possible supercell thunderstorms?

In the space below, give a reason for your answer.

17. Based on the reports given on page 120, which storm(s) is/are classified as severe?

18. Which arrow in Figure 8 points to the location of the heaviest rain and hail?

19. Which arrow in Figure 8 points to the location of a possible tornado?

APPENDIX A

CONVERSION FORMULAS

LENGTH CONVERSIONS

1 kilometer (km) = 0.62 miles
1 kilometer (km) = 3,281 feet
1 mile = 5,280 feet

TEMPERATURE CONVERSIONS

$^{\circ}C = 5/9(^{\circ}F - 32)$
$^{\circ}F = 9/5^{\circ}C + 32$
$K = ^{\circ}C + 273$

APPENDIX B

DOPPLER RADAR VELOCITY IMAGE

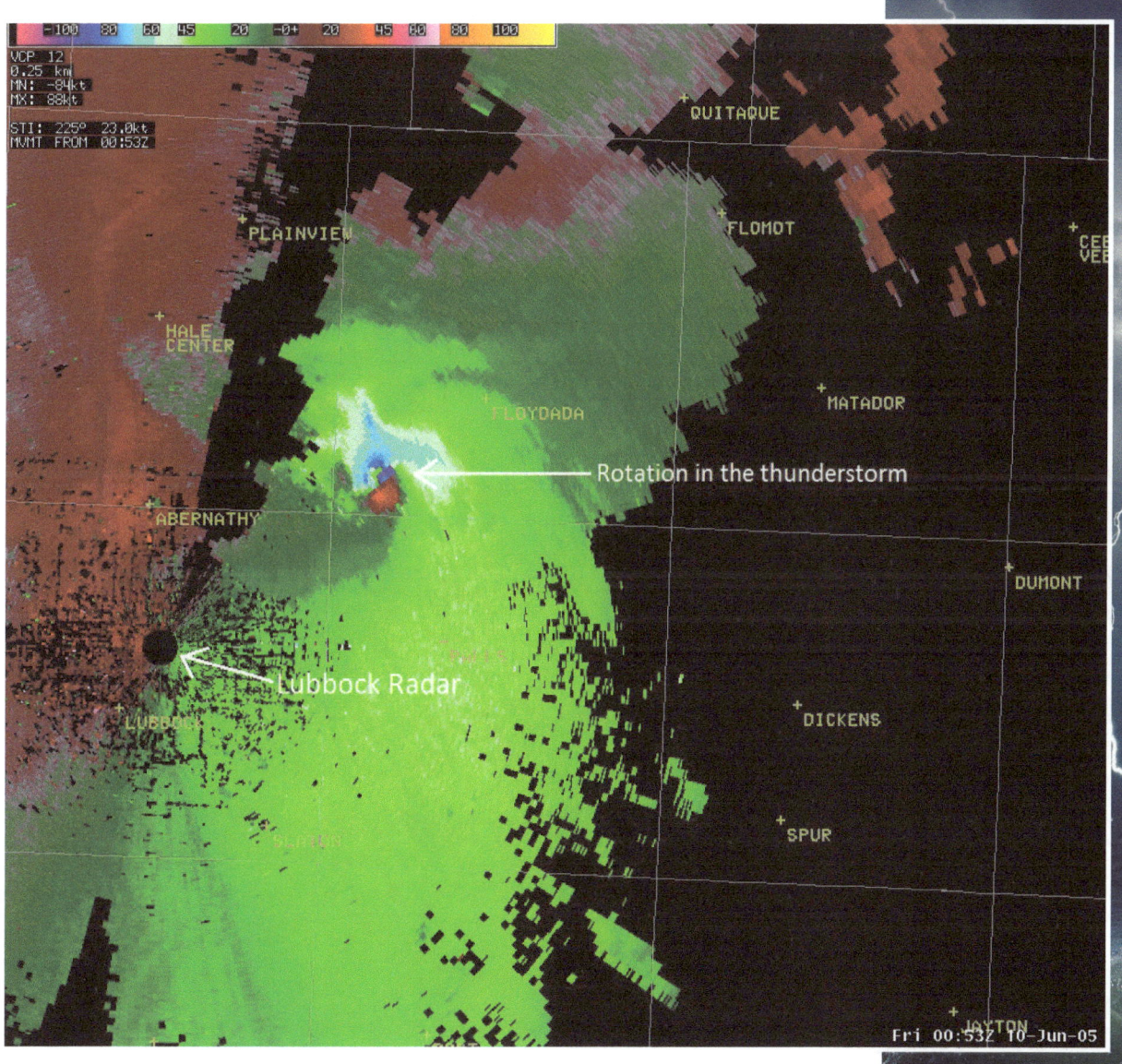

Source: National Weather Service

APPENDIX C

VISIBLE SATELLITE IMAGE

This image was taken at 5:10 pm CDT on the same day used in Lab 8.

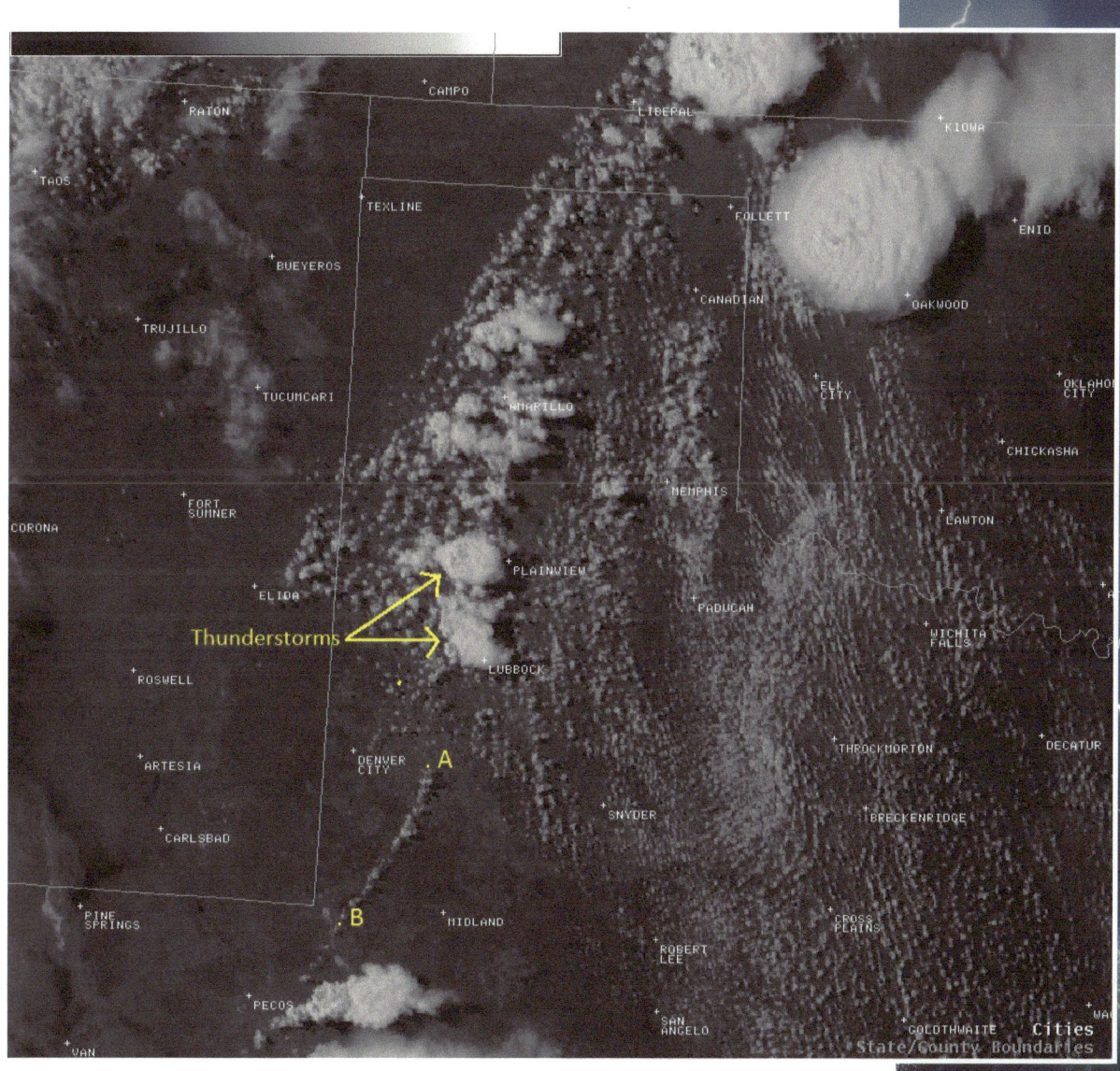

Source: National Weather Service

GLOSSARY

A

Absorption – A process in which radiation is retained by a substance.

Actual mixing ratio – The ratio of the actual mass of water vapor to the mass of dry air. It is used to express the amount of water vapor in the air.

Adiabatic – A process involving a parcel of air in which there is no transfer of heat energy between the parcel and its surrounding environment.

Air mass – A large body of air with similar horizontal temperature and moisture properties.

Anticyclone – A large, high pressure system around which air flows in a clockwise sense in the Northern Hemisphere.

C

Condensation – A change of phase from gas to liquid.

Conduction – The process of transferring energy from molecule to molecule within a substance or between substances that are in direct contact.

Convection – In meteorology, the vertical transfer of energy by the movement of air.

Convergence – The net horizontal inflow of air into an area.

Cyclone – A large low pressure system around which air flows in a counterclockwise sense in the Northern Hemisphere.

D

Density – Mass per unit volume.

Dew point temperature – Also referred to as just dew point. The temperature to which air must be cooled at constant pressure for saturation to occur. It is an indicator of the actual water vapor content.

Divergence – The net horizontal outflow of air away from an area.

Dryline – A boundary separating warm, moist air from warm, dry air.

Dry adiabatic lapse rate – The rate at which an unsaturated parcel of air cools as it rises or warms as it sinks. It is a constant equal to 1°C/100 meters or 10°C/km.

E

Evaporation – A change of phase from liquid to gas.

F

Front – A boundary or transition zone between two different air masses.

H

Humidity – A general term for the amount of water vapor in the air.

I

Insolation – The amount of incoming solar energy over the daylight hours at some location.

Instability – A term associated with unstable parcels of air that rise due to positive buoyancy.

Inversion – An increase of temperature with height.

Isobar – A line connecting points of equal pressure.

Isopleth – A generic term for a line connecting equal values of a given variable on a map.

Isotherm – A line connecting points of equal temperature.

J

Jet stream – A relatively narrow band of rapidly moving air (strong wind speeds) located near the tropopause.

K

Kinetic energy – The energy an object possesses because of its motion.

Knot – A unit of wind speed equal to one nautical mile per hour. 1 knot = 1.15 statute miles per hour.

L

Lapse rate – The rate at which temperature decreases with height.

Level of Free Convection (LFC) – The altitude at which a rising, stable parcel of air becomes neutrally stable. Above this altitude, the saturated parcel is unstable and freely rises due to buoyancy.

Lifting Condensation Level (LCL) – The altitude at which a rising parcel of unsaturated air becomes saturated.

Longwave radiation – A term used to describe the infrared radiation emitted by the Earth and atmosphere.

M

Millibar (mb) – A unit of atmospheric pressure.

Mixing ratio – The ratio of the mass of water vapor to the mass of dry air. (See actual mixing ratio and saturation mixing ratio.)

P

Parcel of air – An imaginary representative sample of air about the size of a large bubble or balloon used to explain atmospheric processes such as cloud formation and atmospheric stability.

Pressure – Force per unit area. In the atmosphere, a measure of the weight of air above a given location.

Pressure gradient – The change (or difference) in pressure over some distance.

R

Radiation – The transfer of energy through space or a medium in the form of electromagnetic waves.

Radiative equilibrium temperature – The temperature at which the rate of absorption of solar radiation equals the rate of emission of longwave radiation.

Radiosonde – A device carried upward from the surface by a balloon that measures temperature, pressure, and humidity at different altitudes through the troposphere into the stratosphere. Wind direction and speed can be derived by tracking the device.

Relative humidity – The ratio of the actual water vapor content to the maximum possible water vapor content. It is expressed as a percentage.

S

Saturated adiabatic lapse rate – The rate at which the temperature in a saturated parcel of air changes during vertical motion. It is sometimes called the moist adiabatic lapse rate.

Saturation – A condition in the atmosphere in which air contains the maximum amount of water vapor possible for a given temperature.

Saturation mixing ratio – The ratio of the mass of water vapor to the mass of dry air when the air is saturated. It is used to express the maximum possible amount of water vapor in the air for a given temperature.

Scattering – A process by which radiant energy is redirected in different directions by gas molecules and particles in the atmosphere.

Sea level pressure – Atmospheric pressure at mean sea level. Surface pressure is converted to sea level pressure to remove the elevation differences among locations.

Shortwave radiation – A term used to describe energy emitted by the Sun. It includes visible and ultraviolet wavelengths.

Solar angle – The angle between the horizon and the Sun.

Solar declination – The latitude at which the Sun is directly overhead at solar noon.

Source region – A large area of land or water over which air masses form.

Stability – A condition of the atmosphere that either favors or resists vertical air motion. Parcels of air can be stable, unstable, or neutral. Stability largely determines the type of cloud that forms.

Station model – A format used to plot weather observations from a single station on a map.

Supercell – A type of thunderstorm characterized by a rotating updraft. Supercell storms can exist for more than an hour and can produce large hail and in some cases, tornadoes.

Surface pressure – The pressure measured at a given location on Earth due to the weight of air above that location. Also called station pressure.

T

Temperature – A measure of the average kinetic energy of the molecules of a substance.

Tropopause – The boundary between the troposphere and stratosphere.

V

Vertical wind shear – The change of wind direction and/or wind speed with height.

W

Wavelength – The distance between two subsequent wave crests.

Wind – Air in motion relative to the Earth's surface.

Z

Zenith angle – The angle between a point directly overhead and the Sun.